Eclipse!

The **What**, **Where**, **When**, **Why**, and **How** Guide to Watching Solar and Lunar Eclipses

Philip S. Harrington

John Wiley & Sons, Inc.

New York • Chichester • Weinheim • Brisbane • Singapore • Toronto

For my mother-in-law, Helen Hunt, and grandparents-in-law,
John and Anna Musick, with love and thanks

This text is printed on acid-free paper.

Copyright © 1997 by Philip S. Harrington
Published by John Wiley & Sons, Inc.

This publication is designed to provide accurate and authoritative information
in regard to the subject matter covered. It is sold with the understanding that the
publisher is not engaged in rendering legal, accounting, or other professional
services. If legal advice or other expert assistance is required, the services of a
competent professional person should be sought.

Library of Congress Cataloging-in-Publication Data

Harrington, Philip S.
    Eclipse! : the what, where, when, why, and how guide to watching solar and
lunar eclipses / Philip S. Harrington.
       p.   cm.
    Includes bibliographical references and index.
    ISBN 0-471-12795-7 (pbk. : alk. paper)
    1. Solar eclipses—Observers' manuals. 2. Lunar eclipses—
Observers' manuals. I. Title.
QB541.H35 1997
523.7'8—dc21   96-29777

Printed in the United States of America

10 9 8 7 6 5 4 3 2 1

Contents

Preface

Amongst all the wonders of all the wonderful sciences there is no science which deals with such a gorgeous spectacle as is exhibited by the queen of sciences, astronomy, at the moment when the earth is gradually shrouded in darkness and when around the smiling orb of day there appears the matchless crown of the corona. Nor can any science duplicate the wonderful precision shown by the work of the astronomer in his capacity to predict hundreds of years in advance the exact hour and minute at which an eclipse will take place and the locality where such an eclipse will be visible.

So wrote S. A. Mitchell in *Eclipses of the Sun* (1923). Back then, few people followed eclipses with the fervor and dedication that we see today. Enthusiasts think nothing of jetting around the world just to witness the few brief moments of a total solar eclipse. And with good reason, for all who behold the majesty of totality will give impassioned testimony to its unbridled glory. A total solar eclipse is the most beautiful and emotionally charged celestial event of all.

Annular, or "ring of fire," solar eclipses also attract wide attention, as do partial eclipses. And the more common lunar eclipses, which have always taken a backseat to solar eclipses, bring with them a unique beauty that may be enjoyed by both veteran eclipse watchers and apprentice sky gazers alike. Personally, I always look forward to a total lunar eclipse with great anticipation, perhaps because it was the eclipse of April 1968 that first pulled me into this fascinating hobby and science of astronomy. And I have been hooked ever since!

The purpose of the book you hold before you is simple. I want to spread some of my love and enthusiasm for solar and lunar eclipses. Inside, you will find chapters that detail the mechanics of solar and lunar eclipses, that tell you what equipment (if any) it takes to view and photograph eclipses, and that discuss what to look for during each type of eclipse. The book culminates with a pair of chapters that detail all solar and lunar eclipses set to occur between the years 1998 and 2017. Believe me, there are some exciting times ahead for both solar- and lunar-eclipse lovers.

Acknowledgments

Trying to write a book on any technical subject is always a difficult task. Indeed, it might well be impossible—for me, anyway—were it not for the help and contributions from many individuals. These folks have contributed their time and their talent to helping make this book a success. My heartfelt thanks to each of them.

First, to the photographers. What would a book on eclipses be, without some striking photographs? One look through the pages to come and I think you will agree that this book has some of the finest ever taken by amateur astronomers. For these I must thank Steve Albers, John Davis, Daniel Fischer, Akira Fujii, Tim Kelley, Brian Kennedy, Jack Newton, Ernest Piini, Frans Pyck, Spencer Rackley IV, Richard Sanderson, Craig Small, Byron Soulsby, Sam Storch, Bernie Volz, and Paul Whitmarsh.

Many people waded through the early stages of this book. It was these proofreaders' suggestions, comments, and criticisms that helped to mold and steer the book's content into its final form. For giving their honest, sometimes harsh (but always deserved) criticism, I wish to thank Daniel Fischer, Daniel Green, Dave Kratz, Jack Megas, Ernie Piini, Patrick Poitevin, Richard Sanderson, and Craig Small.

The data tables throughout chapters 7 and 8 have been generated by and checked with various software programs, most notably Christian Nuesch's *Lunar Eclipse* and Matthew Merrill's *Solar Eclipse*. My thanks to them both. The maps in the two chapters are based on plots made with *Eclipse Complete* from Zephyr Services. All of these programs are available from sources listed in Appendix A.

Thanks to all those who contributed some wonderful anecdotes and eclipse stories, many of which have been woven into various chapters of the book, as well as to Jay Anderson, Fred Espenak, and Joe Rao for fielding my many questions. My gratitude also to Kate Bradford, my editor at Wiley, who has always been very encouraging throughout the book's gestation period.

None of the photographs, maps, or words would have made their way onto these pages were it not for the love and support of my wife, Wendy, and daughter, Helen. They have tolerated me—indeed, bolstered me—

throughout the more than two years it has taken to bring this book to print, and I love them dearly for their boundless love.

Finally, I want to thank you, the reader, for picking up this book. Viewing eclipses is one of the most inspirational aspects of astronomy, and I truly hope you find these pages to be full of information and excitement. I would like to hear from you, too. If you have comments or questions, don't hesitate to write to me in care of the publisher, John Wiley & Sons, 605 Third Avenue, New York, NY 10158. If you prefer, you may send electronic mail to me directly at either psh@juno.com, pharrington@compuserve.com, or starware1@aol.com.

1 The Dynamic Duo

Gossamer clouds of glowing hydrogen and clusters of colorful stars. Alien planets enveloped in noxious atmospheres and airless worlds pitted with innumerable craters. Frigid comets with long, graceful tails, and infinitely distant galaxies. These are just some of the wondrous sights that the sky holds in store for stargazers. All are unique, all are special. But of all there is to enjoy in our universe, none has drawn more attention, struck greater fear, or captured the hearts and souls of stargazers more than eclipses of the sun and moon.

Those who have witnessed their beauty firsthand describe total solar eclipses as the most awe-inspiring events that nature has to offer. People travel great distances just to witness the stark beauty of the solar corona, chromosphere, and prominences—all features of the Sun that are normally hidden by the intensity of the Sun's brilliant surface, the photosphere.

Lunar eclipses can also be powerfully moving events. As the Moon slips into the shadow of the Earth, it will frequently take on a colorful, ruddy tint that many observers have compared to the view of the red planet Mars through telescopes. The view can be especially memorable if the Moon is nestled within a star-filled backdrop. And unlike a solar eclipse, whose maximum phase is visible over a comparatively small portion of the Earth's surface, a lunar eclipse appears exactly the same from anywhere on the night side of the Earth. This convenience affords many more people the opportunity to enjoy an eclipse's beauty without leaving their homes.

Our fascination with these captivating celestial events is nothing new. Solar and lunar eclipses have attracted a great deal of attention for as long as humans have looked at the heavens.

1

A BRIEF HISTORY LESSON

Throughout human history, the night sky was for the most part considered to be serene and never-changing. Our ancestors took great solace in that feeling of permanence. Although every once in a while an unexplainable event would pass over their heads, such as a shooting star, or meteor, flashing briefly across the heavens, or perhaps simply a "wandering star" (which is the literal meaning of the word *planet*) moving slowly against the backdrop of fixed stars, they usually trusted that these were benign acts of the gods.

But occasionally sky events took place that struck unbridled fear in people's souls. Perhaps a "hairy star," or *comet*, would grace the skies. Until relatively recent times, these were believed to be omens of evil. But for many, the most feared events occurred when either of the two greatest sky gods, the Sun and Moon, were in peril—during an eclipse. Most ancient cultures saw eclipses as portents of evil. Some interpreted eclipses as signs that something or someone was trying to consume or steal these vital sky entities; others viewed them (or at least saw solar eclipses) as the Sun and Moon battling each other. Entire civilizations would drop everything when an eclipse occurred, doing everything in their power to scare away the awful creature bent on destroying *their* gods.

Perhaps as far back as 2800 B.C., the ancient Chinese saw a rhythm in the occurrence of solar eclipses, although they believed a ferocious dragon was devouring the Sun! One frequently recounted tale comes from the Hsia Dynasty. Fearing that the wrath of the cosmos would influence earthly events, the dynasty's fourth emperor, Chung K'ang, relied heavily on two court astrologers, named Hsi and Ho, for celestial forecasts. The story goes that on one fateful day Hsi and Ho failed to predict the occurrence of a solar eclipse. When the emperor confronted the astrologers with this grievous error, he found them both drunk on wine. He became so infuriated by their irresponsibility that he ordered them executed!

The Chinese were not alone in viewing solar eclipses with fear. Diverse cultures in Europe, Africa, Asia, and the Americas believed that a solar eclipse was caused by a terrible monster eating the Sun. The ancient Norse tribes thought that an enormous wolf, named Sköll, gobbled up the Sun during an eclipse, while the tribespeople living in Mongolia and eastern Siberia felt that they were caused by Alkha, another creature who had an insatiable appetite for the Sun. Legend had it that Alkha was beheaded by the gods, but even this did not stop the severed head from searching the heavens for the Sun!

Many civilizations decided that the best way to vanquish the "demon" that was consuming their Sun was to band together and make as much noise as possible to scare it away. At the first sign of an eclipse, everyone would immediately gather in the center of town to bang drums, and shout and scream as loudly as they could. It must have worked; the Sun returned every time!

Solar eclipses have even altered the course of human history. In 585 B.C. the Lydians and the Medes were doing battle in what is present-day Turkey. The Greek historian Herodotus recorded that, at the height of a particularly fierce battle, darkness fell upon the land. Apparently the two armies had the good fortune to wage war very near the path of a total solar eclipse! The armies took this as a sign, stopped fighting instantly, and came together to make peace.

Lunar eclipses are not without interesting tales, as well. Many ancient cultures, including the Greek, Chinese, Islamic, and Mayan, had legends that associated lunar eclipses with plagues, earthquakes, and other disasters.

Perhaps the most famous lunar-eclipse anecdote comes from the exploits of Christopher Columbus. Times had become desperate during his fourth voyage to the New World, as an epidemic of shipworms turned his fleet into a collection of sieves. Finally, conditions forced him to beach his sinking armada on Jamaica. The natives provided the castaways with food and shelter, but tension mounted among the shipwrecked crew as the passing weeks turned into months. Finally, some six months later, half of the crew mutinied, attacking the remaining crew, murdering the natives, and stealing their food. Not surprisingly, this put an immediate halt to Columbus's bartering with the natives for additional food.

Columbus knew from an astronomical almanac he had brought along on the voyage that a total lunar eclipse would be seen from Jamaica in just a few days. He also knew that the Jamaicans were terrified by such events. Capitalizing on this, Columbus told the native chiefs that unless they immediately gave his crew food, the angry Christian God would turn the Moon blood red. Sure enough, that night the eclipse went off as predicted. The terrified natives quickly made amends with ample food offerings, and continued to keep the crew well fed until help from home arrived.

THE AGE OF UNDERSTANDING

While many peoples dreaded eclipses, others yearned to understand exactly what was happening. Perhaps the oldest testament to early humans' attempt to understand the universe is Stonehenge, situated on Salisbury Plain in England. Stonehenge, constructed by several cultures between about 2800 B.C. and 1500 B.C., is believed to have been used to measure the motions of the Sun and Moon. Though most authorities agree that none of the cultures who constructed and used Stonehenge could predict exactly when an eclipse would occur, they may have been able to issue warnings of the likelihood of an eclipse on the order of days or even weeks before it happened.

The ancient Greek astronomer Hipparchus attempted to understand eclipses by using them to make scientific observations. Around 130 B.C., from observations of a solar eclipse seen from Hellespont and Alexandria, Hipparchus determined that the Moon was approximately 429,000 kilometers (268,000 miles) away, only about 11 percent more than today's accepted distance.

Just as Hipparchus was anxious to understand how and why eclipses of the Sun and Moon occur, so also were the Chinese. The first-century-B.C. astronomer Liu Hsiang showed that he was one of the first to understand basic eclipse mechanics when he wrote that the Sun is eclipsed because "the Moon hides him as she moves on her way."

Although these early eclipse pioneers showed great genius in their conclusions, well over a millennium passed before much of the human race began to comprehend the workings of eclipses. One of the first western astronomers to record a scientific observation of a total solar eclipse was Johannes Kepler in 1605, although little attention was apparently paid by his contemporaries. More than a century later, Edmund Halley published his account of the 1715 April 22 (OS) total solar eclipse in the *Philosophical Transactions of the Royal Society* in London. He, too, described the sight, though he misinterpreted much of what he saw.

HOW DOES AN ECLIPSE WORK?

Most of our scientific knowledge of solar and lunar eclipses has been gained within the last century and a half. Today we are well aware that a solar eclipse is the result of the Moon coming between the Earth and the Sun, and that a lunar eclipse is the result of the Earth coming between the Sun and the Moon. Why eclipses only happen at certain times is a bit more complicated.

The Moon orbits the Earth once every 27.3 days (what astronomers call the *sidereal month*), and the Earth orbits the Sun in 365.2 days. From these combined motions, it has been found that the Moon takes 29.5 days—about two days longer than the sidereal month—to go through a complete

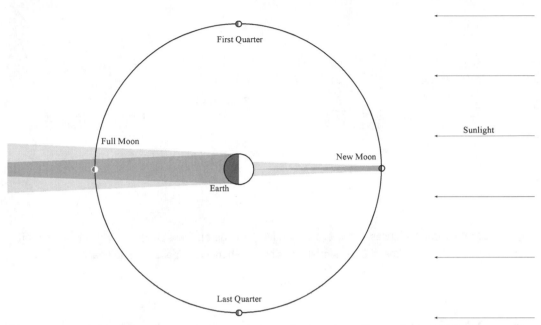

Figure 1.1 The Moon's orbit around the Earth. Note that a solar eclipse, when the Moon passes between Earth and Sun, can only occur during the New Moon phase. A lunar eclipse, when the shadow of the Earth is cast upon the Moon, can only occur during Full Moon.

set of *phases* (New Moon, then on to First Quarter, Full Moon, Last Quarter, and back to New again). This period is called the *synodic month* (figure 1.1), or a *lunation*.

Why isn't an eclipse seen every month? The answer appears in the edge-on view of the Earth-Moon-Sun system shown in figure 1.2. Notice how the Moon's orbit about the Earth is inclined about 5° with respect to the Earth's orbit of the Sun. As a result, the Moon only crosses the Earth's orbital plane (the *ecliptic*) twice every orbit, at points called *nodes*. The Moon is usually above or below the Sun in our sky at New Moon, and misses the Earth's shadow at Full. Only on the comparatively rare occasions when the Moon passes near a node at the New and Full phases can eclipses take place. If the Moon crosses the Earth's orbit from south to north (i.e., from below the plane to above the plane as seen from the northern hemisphere), it is referred to as the *ascending node* (abbreviated by the symbol ☊). *Descending node* refers to the passage from north to south (above the plane to below), and is symbolized by Ω in the figure.

The nodes gradually shift location along the ecliptic as both the Earth and the Sun play a game of celestial tug-of-war called the *regression of the nodes,* with the Moon caught in the middle (figure 1.3). This drifting of the

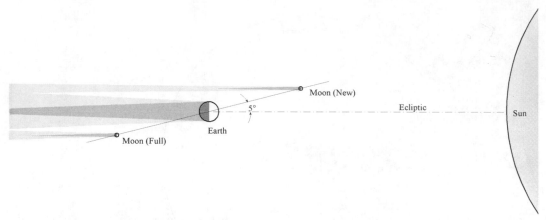

Figure 1.2 Why don't eclipses occur every month? An edge-on view shows how the Moon's orbit is tilted with respect to the plane of Earth's orbit, causing the shadows to pass out into space.

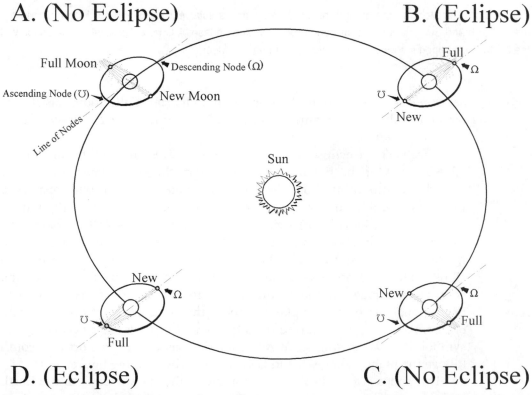

Figure 1.3 A diagram showing an eclipse year. Only when the ascending or descending nodes line up can eclipses occur. Notice how they line up twice during the cycle, each creating two (or more) eclipses.

nodes will realign the New Moon and Full Moon phases with the Sun every 173.3 days, a period referred to as an *eclipse season*. Therefore, each calendar year sees at least two eclipse seasons when solar and lunar eclipses can occur. Two eclipse seasons make up an *eclipse year*, or 346.6 days.

Notice how an eclipse year is shorter than a calendar year by 18.6 days. This difference gives some calendar years not two, but three, four, or even five solar eclipses! There may also be up to five lunar eclipses, but the combined number of lunar and solar eclipses will never exceed seven.

Eclipses typically occur in pairs, with a solar eclipse immediately preceding and/or following (by about 15 days) a lunar eclipse. Imagine this busy scenario. If, at the January New Moon, the Moon just grazes part of the solar disk (producing a partial eclipse of the Sun), it is possible to have a second partial solar eclipse 29.5 days later, at the next New Moon, with the chance of a lunar eclipse in between. The same set of circumstances may occur when the New Moon and the nodes line up again in June and/or July. Finally, the calendar year will end with a fifth solar eclipse in either November or December, also possibly paired with a lunar eclipse. Seven-eclipse years are rare. The last such year, 1982, featured three total lunar and four partial solar eclipses, while the next, 2038, will bring with it four penumbral lunar, one total solar, and two annular solar eclipses.

SOLAR ECLIPSES

As shown in figure 1.1, a solar eclipse is only seen at New Moon, when the Moon moves between Earth and Sun. When all three bodies are aligned, the Moon casts its shadow across a portion of our world's surface, blocking some or all of the sunlight from reaching the affected region. Just how much of the Sun will be hidden depends on where the observer stands, relative to the Moon's shadow.

Again referring to figure 1.1, you will see that there are two parts to the Moon's shadow, just as there are with all shadows. The dark, central, cone-shaped region shown on the figure is called the *umbra*, while the lighter gray, fan-shaped area is the *penumbra*. If an observer is located outside of the penumbra, then no eclipse will be seen. Those who are positioned within the penumbra will see a portion of the Sun covered by the Moon. The farther an observer is within the penumbra, the greater the percentage of Sun covered by the Moon. Those situated within the umbra will see a *total solar eclipse* (figure 1.4). Since both Earth and Moon are moving, the umbra will trace a line along the Earth's surface during an eclipse, creating

Figure 1.4 A total eclipse of the Sun, the heavens' most glorious celestial sight. Photo by Ernie Piini. (Total eclipse of 1983 June 11 from Jogjakarta, Java; 88-mm f/6.8 refractor, 1-second exposure on Ektachrome 200 slide film.)

a *central path of totality*. The region affected by a partial eclipse is usually bow-shaped, owing to the Earth and Moon's movement, as well as the Earth's curvature.

The Total Solar Eclipse

For a few precious moments the Moon will completely cover the Sun entirely along the central path of a total solar eclipse. During totality, the *photosphere*, the blindingly bright surface of the Sun that is visible on any sunny day, is hidden from view, allowing other, normally invisible features of the Sun to be seen. Surrounding the photosphere is a thin, deep-red layer of the Sun called the *chromosphere*. Measuring only a few thousand kilometers thick, the chromosphere can usually be seen for only scant seconds at the beginning and end of totality. Protruding from behind the Moon's silhouette are the glorious, flamelike *prominences*, stretching for thousands of kilometers into space. Finally, encircling the eclipsed Sun and extending for several times the Sun's diameter, is the pearly white *corona*. As we will explore in chapter 3, the appearances of both the prominences and corona vary from eclipse to eclipse.

Timing and location are everything when viewing a total solar eclipse. By the time it reaches Earth, the Moon's umbra is only 270 kilometers (170 miles) across at its widest, even though it can travel a third of the way around the Earth in a matter of a few hours (figure 1.5). Therefore, the chance that the umbral shadow will pass over any one particular spot on the Earth is slim. Oddsmakers say that any one given point on the Earth can expect to see a total solar eclipse on average only once in 360 years.

The Sun measures 1,392,000 kilometers (864,900 miles) in diameter, while the Moon is a comparatively puny 3,476 kilometers (2,160 miles) across. That works out to be a ratio of approximately 400 to 1; that is, the

Figure 1.5 The shadow of the Moon projected onto the Earth during the 1991 July 11 total solar eclipse. This montage of weather-satellite images shows the shadow at several discrete points along the path of totality. Courtesy of Dr. William Emery and Timothy D. Kelley, Colorado Center for Astrodynamic Research, University of Colorado at Boulder.

Sun is about 400 times larger in diameter than the Moon. At the same time, the Sun is about 149,600,000 kilometers (93,000,000 miles) from Earth, while the Moon is right next door at 384,500 kilometers (240,000 miles) away. Call it divine intervention or just dumb luck, but that ratio also works out to be about 400 to 1. As a result, the Moon and Sun each appear the same size in our sky—about half a degree.

This near-perfect "fit," with the Moon just covering the Sun's brilliant surface while still exposing our star's normally invisible chromosphere and corona, is critical to the majesty of a total solar eclipse. If the Moon appeared noticeably larger than the Sun in our sky, these characteristics would be blocked from view; if it were smaller, then the bright surface of the Sun would never be fully covered, leaving them lost in the glare. (It turns out that of all the planets and satellites in our Solar System, Earth is the only world that enjoys this situation. On all of the other planets, their satellites are either too small or too large to cover the Sun so perfectly.)

The Moon's orbit around the Earth is not circular, but rather oval or elliptical. At its closest point (called *perigee*), the Moon is 356,000 kilometers (221,000 miles) away, while at its farthest (*apogee*), the Moon is 407,000 kilometers (253,000 miles) distant. Likewise, Earth's elliptical orbit of the Sun brings it as close as 147,100,000 kilometers (91,452,000 miles) at *perihelion* (the point where Earth is closest to the Sun), and as far as 152,102,000 kilometers (94,562,000 miles) at *aphelion* (the point where Earth is farthest from the Sun). As a result, the apparent sizes of the Moon and Sun vary slightly in our sky (see figure 1.6 for an example of the Moon's apparent size change). The greater the Moon-to-Sun size ratio, the longer an eclipse's duration. At its longest, totality can last 7 minutes, 31 seconds, but eclipses this long are exceedingly rare. Totality during the solar eclipse of 2150 June 25 will last 7 minutes, 14 seconds, longer than any total solar eclipse since the ninth century A.D. Usually the period of totality is shorter than five minutes.

Other Solar Eclipses

When the Moon appears smaller than the Sun as it passes centrally across the solar disk, a bright ring, or *annulus,* of sunlight remains visible at greatest eclipse. These are called *annular eclipses* (figure 1.7). Because the blinding photosphere is never fully covered by the Moon, the chromosphere, corona, and prominences (discussed in more detail in chapter 3) usually remain hidden from view. Instead, viewers see a strange sort of celestial doughnut in place of the Sun. Though they do not attract the wide following of total solar eclipses, annulars are still spectacular in their own right.

Then there are those eclipses that aren't quite total, yet aren't fully annular either. These hybrid events are called *annular-total solar eclipses*

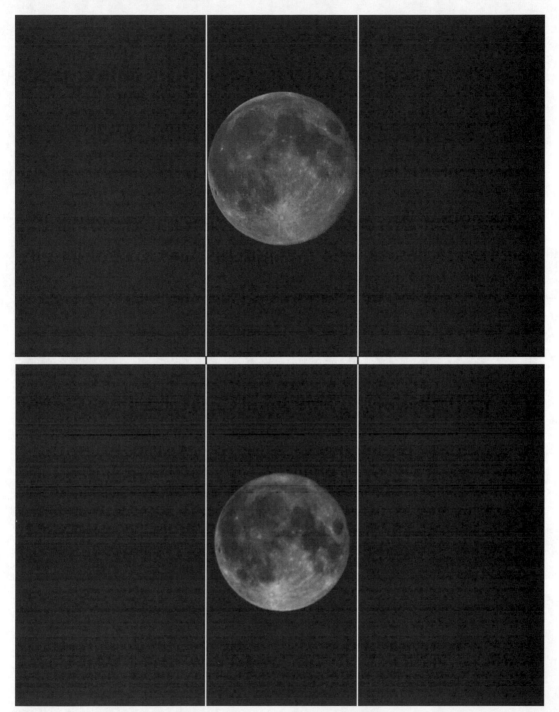

Figure 1.6 One factor affecting the length of an eclipse is the apparent size of the Moon. Compare these two views taken through the same 100-mm f/10 refracting telescope. The view on the top shows the Moon at perigee (closest to Earth), while on the bottom, we see the Moon at apogee (farthest from Earth). Photos by the author.

Figure 1.7 The annular eclipse of 1994 May 10. Photo by the author. (100-mm f/10 refractor with a neutral-density 5 solar filter, ⅟₆₀th-second on Kodachrome 64 slide film.)

(figure 1.8). At the extremes of the eclipse's central path, the Moon appears too small to mask the Sun entirely, and therefore yields an annular eclipse. But because of the Earth's curvature, the Moon's apparent size increases just enough near the point of greatest eclipse to block all of the solar disk, producing a very short total eclipse. Such events are rare, with only two occurring in the twenty-year span of this book.

Finally, there are times when only the lunar penumbra touches the Earth, the umbra casting off into space and missing our world entirely. These circumstances lead to a *partial solar eclipse* (figure 1.9). As with other solar eclipses, the percentage of the Sun eclipsed will vary depending on the observer's location, but regardless of his or her position, the Sun will only be eclipsed partially.

Stages of a Solar Eclipse

Precise moments of key events during a solar eclipse are referred to as *contacts*, and are specified by number: first, second, third, and fourth (figure

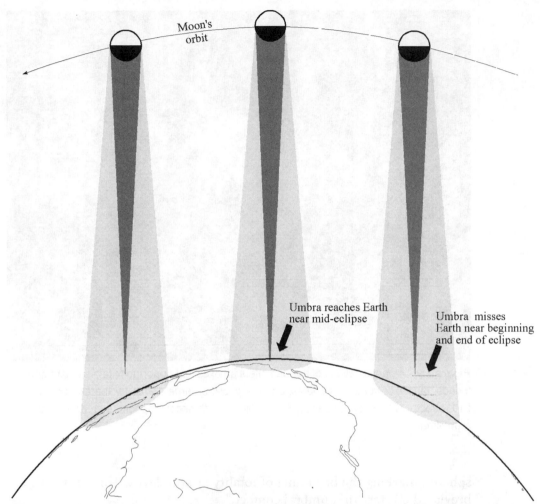

Figure 1.8 The rarest type of central solar eclipse: an annular-total eclipse. The Moon's umbra just reaches Earth's surface near the middle of the eclipse path, but, because of the curvature of the Earth, doesn't quite make it to the Earth near the ends of the track.

1.10). A solar eclipse officially gets under way at a given observing site at first contact, when the Moon first begins to slip in front of the Sun. This point is abbreviated P1, for "first penumbral contact." Keep in mind that, during the partial phases, we are standing in the Moon's penumbra; observers will not see the Moon's umbra until totality.

Second contact, marking the beginning of a central eclipse, will only be seen along the eclipse's central path. This is the instant when the Moon's eastern edge (left edge as seen from the northern hemisphere) first touches the Sun's eastern edge (also the left edge as seen from the northern hemi-

Figure 1.9   A partial eclipse of the Sun offers an excellent opportunity to "sell" astronomy to friends, family, coworkers, students, and the public at large. Photo by the author. (100-mm f/10 refractor with a neutral-density 5 solar filter, ¹/₁₂₅th-second exposure on Kodachrome 64 slide film.)

sphere), marking the beginning of totality or annularity. This contact is abbreviated U1, for "first umbral contact."

Totality or annularity ends when the Moon's western (right) edge leaves the Sun's western (right) edge. This point is abbreviated U2. Midway between U1 and U2 is the point of maximum eclipse.

Finally, last contact (abbreviated P2) signals the end of the partial phases.

Eclipse Magnitude

When astronomers want to define exactly how much of the Sun's *diameter* will appear covered at an eclipse's maximum phase or from a particular vantage point on Earth's surface, they speak of that eclipse's *magnitude*. A magnitude of 1 or greater indicates that the eclipse will be total, while lesser values tell observers that some Sun will remain visible at greatest ob-

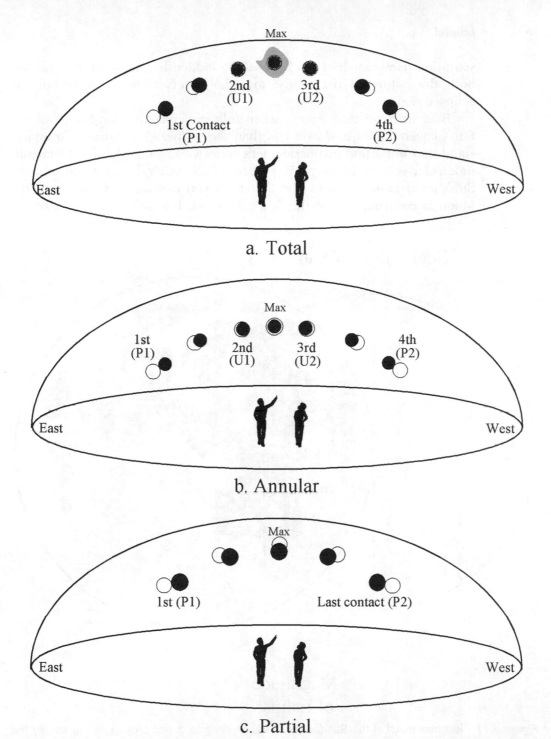

Figure 1.10 The times when major events occur during a solar eclipse are described according to shadow "contacts." The top view (a) shows the contact points of a total solar eclipse. The middle view (b) shows the contact points of an annular eclipse. Note how a portion of the Sun remains visible at both U1 and U2, and even at maximum. The bottom diagram (c) depicts a partial solar eclipse, in which, since the Moon is never fully superimposed over the Sun's face, there are only two contacts, first (P1) and last (P2).

scuration. How much of the Sun will be hidden by the Moon depends on both the nature of the eclipse and the observer's location within the eclipsed region.

Bear in mind that, except when fully covered, the percentage of the Sun's covered area is always less than the eclipse's magnitude. To understand that important distinction, let's consider an example of a 0.5-magnitude solar eclipse, in which 50 percent of the Sun's *diameter* (but not the Sun's *area*) is obscured by the Moon. As you can see in figure 1.11, the Moon is covering half of the Sun's diameter, but only 40 percent of the

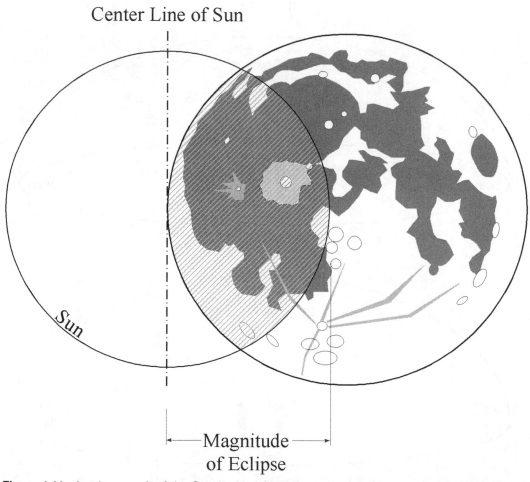

Center Line of Sun

Sun

Magnitude of Eclipse

Figure 1.11 Just how much of the Sun (or Moon) will be covered during an eclipse? To specify this value, astronomers refer to an eclipse's *magnitude.* In the case of a solar eclipse, this is a measure of just how much of the Sun's diameter (not its area) will be covered by the Moon. Similarly, the magnitude of a lunar eclipse specifies how much of the Moon's diameter will be immersed in the Earth's shadow.

Table 1.1 **Eclipse Magnitudes Versus Obscuration**

Magnitude	Obscuration
0.10	0.04
0.20	0.11
0.30	0.19
0.40	0.29
0.50	0.40
0.70	0.63
0.90	0.89
1.00	1.00

Sun's *area.* Table 1.1 compares eclipse magnitudes to disk obscuration (the area of the Sun covered by the Moon). The results are based on a computer program created by Ralph Merletti of Lakewood, Colorado, and published in *Sky & Telescope.*[1]

Eclipse Cycles

Earlier in this chapter, we saw how a solar eclipse can only occur when the plane of the Moon's orbit crosses that of the Earth at the New Moon phase. How often that happens was of great interest to many ancient civilizations, just as it is to us today. The Greeks saw that there was a relationship between the periodicity of the Moon's synodic month and the eclipse year. They calculated that 223 synodic months equals nearly the same period of time as 19 eclipse years: 6,585 days, or 18 years, 11 days. This relationship has since become known as the *Saros.*

You might almost think of a Saros as a family of eclipses. Any two solar eclipses separated by that 18-year, 11-day time span generally will share many common characteristics, such as duration, type (total, annular, or partial), length and shape of the eclipse path, and so on. To illustrate this, consider the remarkably long eclipse of 1973 June 30, which featured 7 minutes, 4 seconds of totality. Adding 18 years, 11 days to that date brings us to 1991 July 11, when totality lasted a maximum of 6 minutes, 54 seconds. Both were in the same Saros, or eclipse lineage, and shared many common traits. The next eclipse in their Saros will occur on 2009 July 22, when totality will last a maximum of 6 minutes, 39 seconds.

Because the synodic and sidereal months and the eclipse year are not *exact* multiples of one another, characteristics of eclipses within the same Saros will slowly change. First and most obvious, since a Saros is not exactly 18 years long, the dates of the eclipses advance by 11 days.

The difference is not *exactly* 11 days; actually, it works out to be 11.32 days, or 11 days, 8 hours. As a result of that 8-hour span, the Earth will have shifted under the Moon's shadow by a third of a day, or about 120° in longitude. So, while the June 1973 eclipse was visible in Africa, the July 1991 eclipse was seen from the Pacific Ocean and Mexico—slightly north and about 120° west.

Since solar eclipses occur more often than once every 18-plus years, clearly there must be more than one Saros cycle running at any given time. Indeed, if you were to do the mathematics involved, you would find that there are 42 Saros series evolving concurrently! In order to keep one straight from another, each has been numbered sequentially for easy iden-tification, using a system first proposed by astronomer George van den Bergh of the Netherlands in 1955.[2] Eclipses assigned odd-numbered Saros numbers take place at the Moon's ascending node, while even-numbered Saros numbers indicate eclipses that occur during descending nodes. For instance, the solar eclipses of 1973, 1991, and 2009 July 22 are members of Saros #136 and all occur at ascending nodes.

Saros cycles eventually end, only to be replaced by new cycles. While none conclude during the 20-year span of eclipses covered in chapter 7, a new Saros (number 156) begins with the partial eclipse of 2011 July 1.

LUNAR ECLIPSES

Figure 1.1 showed that a lunar eclipse will occur whenever the Moon passes through the Earth's shadow. This can only happen at Full Moon. While a solar eclipse is seen over only a small portion of the day side of Earth, and changes in appearance depending on where an observer is standing relative to the point of maximum eclipse, a lunar eclipse can be viewed from across the entire night side of the Earth (weather permitting, of course), and will appear *exactly the same* for all observers on the Earth's night hemisphere. So, while solar-eclipse chasers may have to travel great distances to be in the center of the umbra during a solar eclipse, lunar-eclipse observers may generally enjoy the event from the comfort of their yards without missing a thing.

As suggested earlier, lunar eclipses also come in three varieties: total, partial, and penumbral. A *total lunar eclipse* (figure 1.12) takes place when the Moon's entire disk is bathed in the Earth's umbra. While the Moon's comparatively small umbra focuses very nearly to a point by the time it reaches us, the Earth's larger umbra spans some 16,000 kilometers (10,000

Figure 1.12   The total lunar eclipse of 1993 November 28. Photo by Richard Sanderson. (150-mm f/12 refractor, 15-second exposure on Ektachrome P800/1600 slide film developed at ISO 1600.)

miles) across at the Moon's distance, better than twice the Moon's diameter. The total phase of a solar eclipse may last only a few minutes, but totality during a lunar eclipse can continue for 90 minutes or more.

Key events during a lunar eclipse are also specified by contact times, although they may, at first, appear a little more complicated than their solar-eclipse counterparts. First penumbral contact occurs when the Moon first enters Earth's penumbra, while first umbral contact marks when the Moon is first shadowed by Earth's darker umbra. First penumbral contact (also abbreviated P1) is difficult to discern visually, while first umbral contact (abbreviated U1) is clear with or without optical aid (however, trying to tell *exactly* when first umbral contact occurs is difficult; see chapter 4 for details). Figure 1.13 represents all of these points graphically.

Second umbral contact (U2) takes place at the beginning of totality, when the Moon is fully eclipsed by the Earth's umbra, and third umbral contact (U3) occurs at totality's end. I think you may realize by now that last umbral contact (U4) marks when the Moon completely leaves Earth's umbra, and second penumbral contact (P2) occurs when the Moon leaves

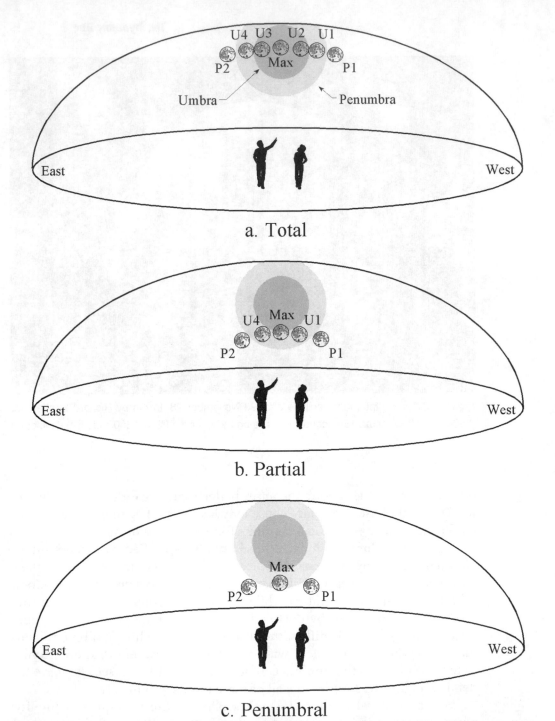

a. Total

b. Partial

c. Penumbral

Figure 1.13 As with solar eclipses, major stages during lunar eclipses are defined by shadow "contacts": a total lunar eclipse (a); a partial lunar eclipse (b), when the Moon never fully enters the Earth's umbral shadow; and a penumbral eclipse (c), when the Moon never enters the darker umbra.

the penumbra. (Note, in the timetables found throughout chapter 8, that partial eclipses do not have U2 and U3 values, since those are reserved for when the Moon fully enters the Earth's umbra. Rather, contact times for partial lunar eclipses are designated only P1, U1, U4, and P2, for first penumbral contact, first umbral contact, last umbral contact, and last penumbral contact, respectively.)

The Moon never completely disappears during a total lunar eclipse. Even at maximum, a small amount of sunlight is bent, or refracted, through our atmosphere and into Earth's shadow. Because of the refractive properties of the atmosphere, light from the blue end of the visible spectrum is scattered, while light from the red end is more readily passed. The result is a reddish cast to the Earth's shadow, the Moon taking on this same crimson coloring during a lunar eclipse.

No two lunar eclipses appear exactly the same. Sometimes the umbra appears a bright red-orange; at other times it appears dark coppery red or even brownish gray. A total lunar eclipse's color is dependent upon the clarity of our planet's upper atmosphere. In general, a vivid eclipse will be seen when the air is free of particulate matter, while a darker total phase will be witnessed when the atmosphere is polluted with foreign particles, such as volcanic aerosols. The lunar eclipse of 1982 December 30 was dimmed because of that year's volcanic eruption of El Chicon in Mexico, and the 1991 eruption of Mount Pinatubo in the Philippines darkened the eclipse of 1992 December 9–10. Neither was as dramatic, however, as the 1963 December 30 eclipse, when the Moon all but vanished from naked-eye view thanks to the eruption of Mount Agung in Bali, earlier that year.

A *partial lunar eclipse* (figure 1.14) occurs when the Moon is oriented such that only a portion of it dips into the Earth's umbra. Depending on the magnitude of the eclipse, the shadowed portion of the lunar surface may appear a dark red or rust color, or simply a charcoal gray, because of the sharp contrast between it and the brilliant part of the Moon that remains outside the umbra. A similar visual effect may be expected during the partial phases before and after the total phase of a total lunar eclipse.

Finally, *penumbral lunar eclipses* (figure 1.15) occur when the Moon only passes through the faint penumbral portion of Earth's shadow. None of the lunar surface is completely shaded by Earth's umbra; instead, observers see only the slightest dimming near the lunar limb closest to the umbra. Indeed, unless at least half of the Moon enters the penumbra, the eclipse may prove undetectable!

Just as a solar eclipse is referred to by its magnitude (that is, how much of the solar disk will be eclipsed), so too is a lunar eclipse specified by a magnitude value. Actually, a lunar eclipse has two magnitude values associated with it: the penumbral magnitude (usually abbreviated PMAG)

Figure 1.14 A partial eclipse of the Moon, as photographed by the author. (200-mm f/7 reflector, $1/60$th-second exposure on Ektachrome 400.)

Figure 1.15 The Moon experiences only a subtle change during the penumbral phase of a lunar eclipse. Photo by the author. (200-mm f/7 reflector, $1/250$th-second exposure on Ektachrome 400.)

and the umbral magnitude (UMAG). The former refers to how much of the Moon's diameter (not its *area,* as you will recall from the previous discussion) will be within the Earth's penumbral shadow, while the latter refers to how much will be in the Earth's umbra at maximum.

Our knowledge of eclipses has certainly come a long way since the days of Hsi and Ho. Now, instead of dreading their occurrence, we anxiously anticipate them. Amateur astronomers and photographers may spend hundreds of hours and thousands of dollars purchasing and readying their equipment just for the few brief moments of an eclipse. But is there a danger to eclipse watching? What equipment do you *really* need? The next chapter addresses some of those concerns.

2 An Eclipse Watcher's Shopping List

Dedicated eclipse watchers anticipate the occurrence of an eclipse with great enthusiasm. Some spend weeks preparing their equipment, making certain that everything will work just right. Others, feeling that a lot of expensive equipment is necessary to view an eclipse, draw up huge shopping lists of things to buy before the event draws near.

Today's eclipse enthusiast can select from a wide variety of instrumentation. The choice is so great that it can be difficult to decide between the absolutely necessary, the useful but not critical, and the useless. Where do you begin? Right here. But before we discuss telescopes, binoculars, and other eclipse-viewing paraphernalia, let's put a few common misconceptions to rest.

I have heard accounts of parents, teachers, principals, and others sequestering children behind locked doors and latched shutters for fear that just being outside during an eclipse will cause them harm. For some reason, people seem to think that the light from an eclipse—*any* eclipse, solar or lunar—takes on a special, almost demonic character. Here's the real story.

First, eclipses of the Moon are perfectly safe to look at for as long as you like. The Moon is illuminated by reflected sunlight, just as the Earth's surface is during the day; gazing at it directly will never damage your vision.

Solar eclipses, on the other hand, do present some danger if viewed incorrectly. If you are unfamiliar with how to view the Sun safely, then please review the next section.

SAFETY FIRST

The Sun holds the dubious distinction of being the only celestial object that may actually harm someone who looks at it directly. What makes looking at the Sun so dangerous? The answer lies in the fact that not only does the Sun radiate visible light, but its photosphere also emits intense infrared (IR) and ultraviolet (UV) radiation. Just as ultraviolet radiation causes sunburn to exposed skin, so too will it damage your eyes' retinas—and at a much faster rate. The human eye need only be exposed to direct sunlight for a few seconds before permanent eye damage, and even blindness, results.

The only way to view the uneclipsed or partially eclipsed Sun safely is either to project or filter the harmful rays of the solar photosphere. During the total phase of a solar eclipse, the Moon does the work for you by completely masking the photosphere. Totality is the only time when it is ever safe to look at the Sun directly and without precaution. All other times require observers to take action beforehand. Fortunately, there are safe and simple ways of looking at the Sun.

Projection

The safest way to view the Sun is to project its image onto some sort of a screen, such as a white piece of paper or cardboard. Projection works equally well with or without a telescope or binoculars.

To watch the progress of a partial solar eclipse with just your eyes, construct a pinhole projector from two pieces of cardboard, as shown in figure 2.1. Punch a small hole, about the diameter of a pencil, in one of the pieces, and glue or tape a piece of white paper on the other. By holding the pieces toward the sky so that sunlight shines through the hole, a tiny image of the solar disk will be cast onto the paper. Although the small image is not as sharp as if it were projected through a telescope, it should be satisfactory for viewing partial phases.

An even better pinhole projector can be made from a long, thin mailing tube (figure 2.2). Punch a small hole into one of the tube's end caps using a nail, and replace the other cap with a thin piece of tissue paper or onionskin paper. Then tilt the tube up toward the sky so that sunlight shines through the hole to cast an image of the Sun onto the paper, like a rear-projection screen. Just make certain that no one tries to look through the paper toward the Sun.

Small telescopes[1] (8-inch [200 mm] aperture and smaller) and binoculars are ideal for projecting a magnified image of the Sun onto a screen

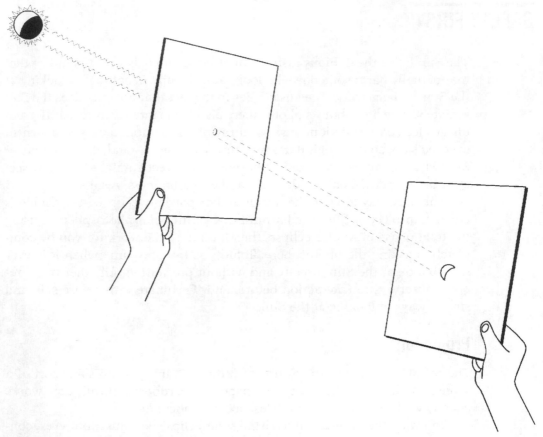

Figure 2.1 Viewing a solar eclipse can be dangerous if not done correctly. The simplest and safest way to watch the progress of a partial or annular eclipse is to project the Sun's image onto a white piece of paper or cardboard, using a pinhole projector.

made from white cardboard (figure 2.3). First, aim the instrument at the Sun. *NEVER* look through the telescope's eyepiece or side-mounted finder scope to do this. Instead, move the instrument up and down, and back and forth, until its shadow on the ground is at its shortest. The Sun should then be in, or at least near, the field of view. (Be sure that no one looks into the eyepiece, especially young children who may be unaware of the danger.)

Once the instrument is centered on the Sun, turn the focusing knob until the projected image appears sharp and clear on the cardboard. The size of the Sun's image can be enlarged by moving the cardboard in and out (the instrument will have to be refocused each time the distance changes). If the image is too small, move the screen away from the eyepiece; if it is too large or faint, reduce the separation.

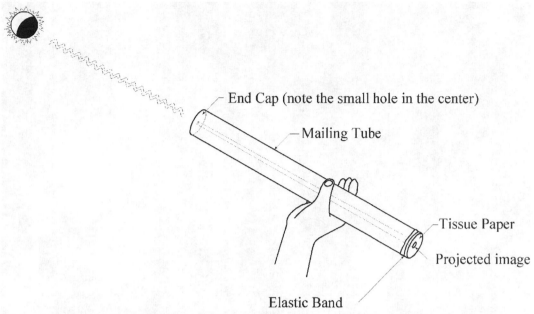

Figure 2.2   Another simple solar-eclipse viewing device can be made with a long mailing tube. Aim the tube toward the Sun (never looking through the tube!) until you see its tiny image projected onto the tissue. Bracing the tube against a tree branch or fence will make supporting it much easier.

To prevent the projected solar image from being washed out, shade it from direct sunlight. Most telescope designs will let you direct the projected image into a cardboard box, like that shown in figure 5.6. Observers trying to project the Sun's image through binoculars or with a straight-through refractor, however, will find the image much improved by first cutting out a cardboard baffle and fitting it securely over the front of the barrel, as shown in figure 2.4.

Before projecting the Sun for an extended time, please heed this warning: The intense heat of the Sun can damage an instrument's delicate optics or soften the glue used to cement an eyepiece or binoculars together. To prevent that from happening, turn your telescope or binoculars away from the Sun for 30 seconds every 5 minutes to let things cool. It is also best to project the Sun with an inexpensive "sacrificial-lamb" eyepiece (this will be discussed later).

Solar Filters

Many amateur astronomers and photographers prefer using special solar filters for viewing the Sun both during a partial eclipse or on any given

Figure 2.3 Telescopes make great solar projectors. Aim the telescope toward the Sun (NEVER looking through it or through the side-mounted finder) and project the image onto a piece of white cardboard. Shading the image, as the author's daughter, Helen, is doing, will make it appear much more vivid.

sunny day. Before chancing your eyesight, however, make certain that you use the proper type of solar filter. Only use filters that are designed for the task and have a filtering factor equivalent to neutral-density 5 (sometimes abbreviated ND-5).

For naked-eye observers, a number-14 (sometimes abbreviated #14) welder's glass makes a safe solar filter, yielding a dark-green image. Your best bet for finding them is in better-stocked welding supply stores, but don't be surprised if they have to be specially ordered. Keep in mind that the comparatively poor optical quality of the glass restricts their use to the naked eye only.

For naked-eye viewing, you can also use two layers of *fully* exposed and developed conventional black-and-white negatives, as the silver in the emulsion of the completely blackened negatives acts to divert the Sun's powerful rays. But take heed; there are so-called "silverless" black-and-white films available today, such as Ilford XP-2. These, as well as all color films, are **not** safe to use as solar filters. Go to your local camera store and

Figure 2.4 Binoculars may also be used to project the Sun onto a piece of white cardboard. Note the shade on the front of the binoculars to help block sunlight from flooding the projected image.

buy a 120-size roll of Kodak Tri-X or Plus-X. Open the package and unroll the film on a sunny day. Have a good time with it. Dance around your yard, wave the film wildly in the sunlight, then take the film back to the camera store and tell them to develop the roll fully as negatives (no need for prints!). Of course, if you have a darkroom, you can do the processing yourself, but make sure you leave the film in the developer for the maximum time. Layer two strips of negatives and hold them across your eyes to watch the Sun. Two layers are equivalent to a neutral-density 5 filter, while three layers equal neutral-density 7.5.

Many department-store telescopes used to come with Sun filters that screw into their eyepieces. If you have one of those filters, *throw it away!* They often crack under the magnified and focused intensity of the Sun. Likewise, do not use photographic neutral-density filters or smoked glass. Both can lead tragically to blindness.

Proper solar filters cover the telescope's full aperture and fit securely over the *front end* of a telescope, camera lens, or binoculars, reducing the Sun's brightness by a factor of more than 99 percent before it enters the optical system (figure 2.5).

Figure 2.5 A front-mounted solar filter is the only safe method for viewing the Sun directly, during either a partial or an annular eclipse, or just on a sunny day.

Sun filters are commercially available in two varieties, either aluminum-coated Mylar or nickel-chromium coated glass. Which type is better? Both diminish the Sun's blinding light to safe levels to yield excellent resolution and contrast.

Aluminum-coated Mylar filters, such as Solar Skreen® by Roger W. Tuthill, Inc. (see Appendix A), are the less expensive of the two. The only real drawback to Mylar filters is that they tend to absorb more of the red end of the spectrum, and therefore produce a blue image of the Sun. While it is easy to overlook the off-color Sun for visual observations, the blue tint rendered by Mylar filters tends to be accentuated in color photographs. Placing an orange filter (referred to in photographic parlance as a number-21 filter) between the filter and the camera will help correct the difference.

Glass solar filters produce a yellowish image. If purchasing a glass solar filter, ask the manufacturer which side of the glass is coated. Better filters are coated on their inner surfaces (the side facing in toward the telescope), as this protects the coating from possibly being scratched during use. Glass filters are available from Thousand Oaks Optical, Orion Telescope Center, and J.M.B., Inc.; consult Appendix A for addresses.

At the upper end of the solar-filter price scale are hydrogen-alpha (Hα) filters. Unlike the glass and Mylar filters described above, which show the Sun across the entire visible spectrum, hydrogen-alpha solar filters display the Sun at one precise wavelength—656 nanometers. Viewing the Sun at that wavelength reveals a turbulence within our star that goes unsuspected through the other, so-called white-light Sun filters. Huge flamelike solar prominences and finely woven, threadlike filaments can be watched and monitored daily with a hydrogen-alpha filter. Once again, Appendix A lists some manufacturers.

All solar filters must be treated with great care, or they will quickly become damaged and unsafe to use. Inspect filters regularly, especially those made of Mylar, for pinholes and irregularities in the coating. This can be done by holding the filter up to a bright light. A small hole can be sealed with a tiny dot of flat black paint without causing the image to suffer. Using a toothpick, dab just a bit of paint over the hole. If, however, the filter is more seriously damaged, it must be replaced immediately.

TELESCOPES AND BINOCULARS

The issue of safety having been addressed, we can move on to optical equipment. What is the best instrument for viewing an eclipse? Unlike other areas of amateur astronomy, where expensive instrumentation may be required, just about any optical instrument (including the naked eye!) will do for watching a solar or lunar eclipse. Your decision should be based on where you intend to view the eclipse. If you will be watching it from your backyard, the choice of viewing instrument might be different from what you would use if you were planning to view a total solar eclipse from, say, Sevaruyo, Bolivia, as many did in November 1994.

For general viewing, especially to appreciate the wondrous sight of the full extent of the Sun's pearly white corona or to enjoy the magnificent view of the eclipsed Moon in a sparkling star field, *nothing* can beat binoculars! They are my single favorite instrument of choice for enjoying an eclipse's aesthetic beauty. Many observers prefer the wide fields of 7×35 or 7×50 glasses, though giant binoculars (e.g., 10×70 and larger) offer a nice compromise between magnification and field of view.

All this is not to say that telescopes do not carry benefits for viewing an eclipse. Even the smallest telescopes are ideal for watching the partial phases, offering close-up perspectives that are missed in binoculars. Photography also proves much more successful through telescopes and telephoto lenses than through binoculars.

If you want to travel, and you may have to, perhaps for great distances, if you want to see most upcoming total and annular solar eclipses, telescope portability is key. While it is possible to transport a large instrument to a remote site, most eclipse watchers prefer portable telescopes and binoculars.

Which telescope is best? Many eclipse watchers, especially total-solar-eclipse chasers, favor compact instruments. The table below lists some of the more popular traveling telescopes.

Unless your observing site is noted for steady sky conditions, smaller instruments (apertures of 20 cm/8 inches or less) are preferred over larger telescopes. Large telescopes, though a must for seeing faint interstellar objects, are more adversely affected by poor "seeing" conditions than smaller instruments, often yielding blurrier images. As you can see from the many wonderful eclipse photographs scattered throughout this book, none were taken with telescopes larger than 20 cm in aperture.

EYEPIECES

If you are viewing through a telescope, each phase of an eclipse may be enjoyed through a wide range of eyepieces and magnifications. Select a low-power, wide-field eyepiece to appreciate the full extent of the eclipsed Sun or Moon, then switch to higher magnifications to take a closer look. Each eyepiece offers a unique perspective.

Table 2.1 **Some Recommended Telescopes for the Traveling Astronomer**

Refractors	AstroPhysics Traveler
	Celestron GP-C102 or GP-C102ED
	Tele Vue Ranger, Pronto, or Genesis SDF
Reflectors	Edmund Astroscan
Schmidt-Cassegrain	Celestron 5 or Celestron 8
	Meade 8-inch LX-10, LX-50, or LX-200
Maksutov telescopes	Celestron C90
	Meade ETX
	Questar 3.5

Your selection of eyepiece should depend heavily on how and what you are looking at. If you will be using a solar filter, or will be observing a lunar eclipse, then select better-quality eyepieces such as orthoscopic or Plössl eyepieces. Eyepieces of lower quality, such as the Kellner and Ramsden designs, are prone to internal reflections and other optical aberrations.

If, however, you plan on projecting a partial solar eclipse through the telescope onto a screen, then reverse this advice and use an inexpensive eyepiece, just in case the Sun's intense heat damages its optics. I'd rather have the Sun's heat melt a $30 Ramsden eyepiece than, for instance, a $300-plus Tele Vue Panoptic!

MORE STUFF

Many eclipse viewers usually bring along a whole laundry list of equipment to watch an eclipse. Here is a short inventory of some of the most popular items.

Tape Recorder

A tape recorder (especially a voice-activated model) is always a useful tool during an eclipse. It can be used to record your instantaneous reactions to the eclipse as it progresses, without having to stop to write things down on paper. (I would still recommend carrying a small pocket notepad and a pen or pencil for note-taking, just in case the tape jams or something else goes wrong.)

As outlined in chapter 4, a small, battery-powered tape recorder is essential during a lunar eclipse when you are logging shadow contacts and lunar occultations of background stars. Try to arrange your observing station so that the tape recorder is close enough to pick up your comments without having to hold the microphone in your hand.

Shortwave Radio

To keep track of your observing program during the excitement of an eclipse, it is critical to know the exact time. A wristwatch may serve the purpose, but many eclipse watchers prefer to bring along a portable shortwave radio that picks up time signals from WWV, the National Bureau of Standards radio station in Boulder, Colorado. Depending on radio-signal propagation, these signals can be heard from just about any point on the globe.

Batteries

With all the talk about tape recorders and radios, we come to the next logical step: batteries. If your tape recorder or radio is battery-powered, be sure to install a fresh set beforehand! That goes for all batteries—still and video cameras, flashlights, anything else that is battery-operated. Be sure to bring along an extra set for crucial equipment, in case the first set fails during the eclipse itself. Rechargeable nickel-cadmium (NiCad) batteries, such as those used to power video cameras, are famous for losing their charges abruptly and without a lot of warning.

White Bedsheet

A white bedsheet serves a couple of purposes when viewing either a solar or lunar eclipse. First, spreading out a sheet under your telescope and camera is a good preventive measure against losing small nuts and bolts in the inevitable pre-eclipse panic. If something drops, it won't fall into the bottomless pit of grass and parking lots; instead, it will come to rest on the sheet.

If you are viewing a total solar eclipse, lay a second white bedsheet alongside your station, to catch a glimpse of shadow bands. As mentioned in chapter 3, these peculiar dark and light ripples are of such low contrast that they are only readily visible against a white or neutral background.

Still More Stuff

Total and annular solar eclipses seem to carry with them their own special needs for a variety of small items that observers often forget to bring along. Here is a list of some things that might prove useful before, during, and after the eclipse.

If the event is going to occur in an especially hot climate, it might be best to cover your cameras and telescopes with a Space Blanket or similar sheet of reflective Mylar. This will help prevent the optics, especially closed-tube instruments such as refractors and Schmidt-Cassegrains, from overheating while waiting for the eclipse to begin. In fact, for the July 1991 total solar eclipse in Baja California, I packed a ten-by-ten-foot picnic table canopy! Sure, I may have looked a little peculiar carrying it on the bus, but was I happy that I had brought it along when the temperature reached 110 degrees!

There are also times when the cold can play havoc with equipment. Mechanical devices (such as camera shutters and telescope clock drives), pliable items (such as film, videotape, and rubber eye cups), and batteries should be kept out of the cold as much as possible. If conditions are especially harsh, try holding more cold-sensitive items inside your coat, where your body's heat will keep things warm until they are needed.

The following table lists some pieces of equipment most commonly needed during solar and lunar eclipses. Not all items are useful during all types of events (there will be little use for a solar filter during a total lunar eclipse, for instance), so abridge and expand the list to suit your own particular needs.

Table 2.2 **Solar- and Lunar-Eclipse Equipment List**

Telescope	____
Binoculars	____
Eyepiece(s)	____
Solar filter	____
Solar projection screen	____
Camera	____
Camera lens	____
Film	____
Cable release (bring a spare)	____
Video camera	____
Videotape	____
Spare batteries	____
Tripod	____
Plastic coffee-can lids	____
(set under tripod feet to keep them from sinking into sand or loose dirt; see chapter 5)	
Tape recorder(s)	____
Space Blanket	____
Sunblock and hat (solar eclipses)	____
Radio (for time signals)	____
Stopwatch	____
Tool kit	____
Thermometer (solar eclipses)	____
White bedsheet(s)	____
Duct tape	____
Elastic bands	____
Logbook	____
Sundry items (food and water, appropriate clothing, etc. Remember that you might be outdoors for a long time)	____

Yes, the well-equipped eclipse chaser goes prepared for just about every contingency. Though it may take some effort and expense to pull all of the equipment and materials together, you'll find that the eclipse will be far more enjoyable (and far less frantic) with a little preparation and foresight.

3 Sun Worshiping

Watching the progress of a solar eclipse, especially a total solar eclipse, can mean different things to different people. Some want to capture its spectacular beauty on film. Others want to perform elaborate scientific experiments on portions of our star that can only be readily observed during the few brief moments of totality. Still others come to see a solar eclipse simply to witness firsthand the sheer excitement and beauty of the Moon's and Sun's coupling.

The drama of the partial phases leading up to a total solar eclipse can be among the most exciting moments of one's astronomical career. Then, suddenly, the Moon completely covers the Sun to reveal the magnificent hidden treasures of our star. All too quickly, however, the Moon continues on its way and the brilliance of our "regular" Sun returns. How quickly the whole thing transpires!

Time is the biggest obstacle for people who view a solar eclipse, especially a total solar eclipse. Most eclipse chasers want to do everything: photograph it, videotape it, look for shadow bands, stars, and planets, watch for the approaching shadow; the list goes on and on. Indeed, some people get so caught up in the moment, taking photographs, fumbling for film, and dropping eyepieces, that they never *see* the eclipse!

Don't make the same mistake that so many have made in the past: trying to do so much during the eclipse that you miss it! That's why this chapter precedes the one on eclipse photography. Forget about photography! Capturing a solar eclipse on film is certainly one aspect of the event, but the *real* thrill of a solar eclipse is seeing it, living it, experiencing for yourself the shock and beauty of the Moon "devouring" the Sun.

Here's an overview of what to expect during an eclipse of the Sun, be it partial, annular, or total. Each of the three following sections is further subdivided with suggested features to look for, and projects for the observer. The projects described here are meant to offer readers some broad alternatives; please do *NOT* interpret this to be a suggested program for a single observer. Instead, select and complete just one or two of these activities; try to do more than that, and you risk missing the eclipse itself!

PARTIAL SOLAR ECLIPSES

Many books all but ignore partial eclipses, saying that they are poor substitutes for "the real thing." Though not as heart-stopping as total or annular events, partial eclipses (figure 3.1) are anything but boring. They occur much more frequently from any particular location on the globe, and can be seen over a wide area. A partial eclipse, whether or not it leads to total-

Figure 3.1 A partial eclipse of the Sun always can afford many people the chance to view the excitement of the Moon crossing the Sun firsthand. Photo by Brian Kennedy. (600-mm telephoto lens at f/8 and a neutral-density 5 solar filter, $1/125$th-second exposure on Kodachrome 64 slide film.)

ity or annularity, offers a wonderful opportunity to experience the magic of astronomy. In addition to capturing the moment on film and videotape, which is discussed in detail in chapter 5, a partial solar eclipse gives amateur astronomers the chance to make some interesting observations about the universe in motion, and the impact that the dimming sunlight has on our changing environment.

Contact Times

The show begins when the Moon takes that first bite out of the Sun's western edge, or *limb*, at First Contact. That first notch tells you that the fun has begun, but, more important, that the universe is working! Think back to those ancient times when eclipses were seen as unpredictable, evil events conjured up by angry gods. Then consider how far our knowledge of celestial mechanics has come, to be able to predict the coming of an eclipse with such amazing accuracy, centuries in advance.

If you have access to a shortwave radio and a stopwatch, you can time the moment when the Moon takes its first nibble out of the Sun. On eclipse day, tune the radio to either WWV or CHU, two of the time-signal radio stations mentioned in chapter 4. With the time signals chiming in the background, start the stopwatch on an exact minute mark. Then turn to your telescope or binoculars, keeping an eye out for the first little notch taken out of the Sun. (While it will be somewhere along the Sun's western limb, its exact position angle will depend on your location.) When you spot it, stop the watch. To figure out the exact time of first contact, add the accrued time to the hour and minute that you started the watch; that's the time of first contact. After the eclipse, go back and compare your observed time with the predicted time for your location. They ought to be within a small fraction of a minute of one another. Do the same for Fourth Contact, as well as for Second and Third Contacts if you are viewing a total or annular eclipse—that is, if you can remember to do it in the excitement of the moment!

The Lunar Profile

Slowly, as the Moon slips across the face of the Sun, study the silhouette of the rugged lunar terrain. With a telescope you can see deep valleys and tall mountains, all painted black against the brilliance of the Sun's photosphere. Using about 100× or higher, you might even try to identify which mountains and valleys are being seen in the Moon's so-called "marginal zone" (along its limb), but you'll need to use a good Moon map first. One of the best is the set of Lunar Quadrant Maps from the University of Arizona's

Lunar and Planetary Laboratory. These maps accurately plot and label thousands of craters down to two miles across, as well as hundreds of other surface features.

Also, keep an eye on any sunspots visible on eclipse day. Watching the slow eastward motion of the irregular lunar limb engulf each spot in its path is fascinating, especially at high magnification. It can be especially fun to follow the Moon as it crosses a large sunspot. First, the spot's lighter surrounding region (confusingly called the *penumbra,* though it has nothing to do with shadows) is struck by the Moon's pitch-black limb, then quickly the sunspot's darker center (called the *umbra,* but again, no shadows are involved) merges into the Moon, and disappears.

If a sunspot or sunspot group happens to lie near the center of the Sun's disk during a solar eclipse, observers can approximate the spot's diameter by using the length of time it takes the Moon to cover it. On average, the Moon will cover the Sun's disk in just under one hour. Because the Sun measures 1,392,000 kilometers (864,900 miles) in diameter, the Moon must pass over 400 kilometers (250 miles) of the Sun's surface every second. Therefore, if it takes, say, ten seconds for the Moon to cover a centralized sunspot, the spot must measure close to 4,000 kilometers (2,500 miles) in diameter. If a spot being measured is far from the center of the solar disk, however, the estimate will not be accurate because of the disk's curvature.

The Environment

Don't stay so glued to the telescope that you forsake the changes going on in your world. If there are any trees in the area, look at the sunlight cast through the branches. The tree's intertwining limbs can act as tiny pinhole projectors to cast a multitude of crescent Suns across the ground (figure 3.2). Even crisscrossing your fingers will work. Placing a white sheet on the ground will help make the crescents more obvious.

If this is a relatively deep eclipse (a magnitude greater than about 0.75), notice the changing contrast of shadows cast on the ground. Use a light meter or photometer to monitor the ever-changing ambient light levels. While the human eye will adapt to the weakened light, a meter can accurately demonstrate the loss of light during an annular eclipse.

Just as ambient light diminishes during the partial phases of an eclipse, so too may the temperature. Little effect other than normal variances will be noticed during minimal partial eclipses, but larger gradients can be observed and felt as the eclipse deepens. Drops in air temperature of 6°C (10°F) or more may be felt during partial eclipses greater than magnitude 0.75.

Figure 3.2 To view a partial solar eclipse safely, you need nothing but a tree's intertwining branches projecting tiny crescent Suns onto a light surface, such as a sidewalk or the side of a house. This photograph of the 1972 July 10 eclipse was taken by the author.

Animals

Wild or domesticated birds and animals frequently react to the onset of a deep-partial, total, or annular eclipse. With the temperature dropping and the light diminishing, many animals are fooled into thinking that evening is upon them. Birds noisily go to roost, and livestock have been known to make their way back to their barns, confused about this unusually short day!

When (if) you first notice the increased activity, take a look at your watch and make a note of the time. How much of the Sun was obscured when the animals first began to react? Be conscious of what is going on around you, not just over your head.

Public Relations

Perhaps the greatest joy of watching a partial eclipse from one's own neighborhood is being able to share the excitement of the event with non-astronomers, including friends, neighbors, relatives, and especially children.

Many astronomy clubs schedule day-long public activities around a partial eclipse, setting up telescopes in parks, shopping malls, and town centers. When properly planned, a club-sponsored eclipse party is an ideal chance for some great exposure. And what club can't use some positive public relations?

Other readers may want to bring the eclipse to work. Here's one success story from Thomas Miller, who brought the May 1994 eclipse to Manhattan's Wall Street financial district. "I work as a securities trader for a bank here in New York City. The trading floor is about the size of a football field, with over 250 people intensely 'whipping around' billions of dollars each day. These people don't leave their desks for anything. Once, when a city fire inspector ordered people off the floor during a fire drill, it almost came to blows! The day of the solar eclipse, I brought a Mylar solar filter to work with me and headed out on the balcony of our building to view the event with my unaided eyes. At first I invited one or two friends to join. They were followed by a few more, and a few more still. The next thing I knew, there was a line of about sixty people waiting patiently to get a view of the eclipse!"

Eclipses are also wonderful for passing along a fascination for the universe to the next generation. For the May 1994 annular eclipse that passed right over Kalamazoo, Michigan, Eric Schreur helped coordinate an observing activity set up by the Kalamazoo Public Museum planetarium and the Kalamazoo Area Math and Science Center, that let students determine the northernmost limit of the line of annularity. More than 500 high school students, working in teams of two, were set up along five different north/south lines, oriented so they were perpendicular to the eclipse path. Each team was given an identical telescope, and told to observe and time the exact occurrence of first and last contact. Afterwards, the data were collected and compared with the predicted contact times. The results indicated that the observed limit line was a little south of the predicted line, though Schreur guesses this was probably because of the limitations of the telescopes used.

That same eclipse had a different effect on Jim Tomney of Baltimore, Maryland, who notes that bringing eclipses to the public is not always easy. "Since I couldn't get to the center line, I thought it would be a fantastic opportunity to share the event with my sons, Matthew and David, and their classes. I spoke to their teachers a few weeks in advance and made plans to bring my telescope and some eclipse visual aids."

Tomney continues, "A little before noon on eclipse day, I got a call at work from one of my sons. 'Dad, my teacher says for you not to bring your telescope,' a disappointed voice told me. I later learned that, in a hastily called meeting that morning, the principal had informed the teachers that, owing to legal considerations, the county school administration had decided no children would be allowed outside during the eclipse. In fact, students were not even allowed near windows!

"Despite my assurances that no one would go blind doing pinhole projection, I was rebuffed. I guess the moral of the story is that in these litigious days, you need a signed permission slip from the parents in advance if you want to show children an eclipse."

A sad testimony to a potentially wonderful educational event, but a lesson learned for future eclipses!

TOTAL SOLAR ECLIPSES

When you are anticipating a total solar eclipse, the partial phases leading up to second contact can seem to drag on forever. But use that time wisely. In addition to taking photographs and shooting videotape, and performing some of the observations mentioned above, use the last minutes before totality to check your equipment. Make sure everything is as it should be, and that it is all within arm's reach.

Most important, take a look around. Pause and feel the wind. Eclipse watchers frequently comment that the wind seems to pick up during the partial phases, only to quiet to a dead calm immediately before and during totality or annularity. Listen to the sounds of the world around you. In short, experience the totality of totality.

Shadow Bands

A minute or so before second contact, tear yourself away from the Sun and look down toward the ground for *shadow bands*, faint ripples of grayish light that wiggle briskly across the landscape. The appearance is akin to the ripples seen at the bottom of a swimming pool on a sunny day.

It is well known that starlight is bent and refracted as it passes through the many swirling warm and cold layers in our atmosphere. Called *scintillation*, this is what causes stars to twinkle. Sunlight is also affected by our turbulent atmosphere, though on any given day the Sun's large diameter and extreme brightness overpowers the effect. But with the Sun sliced to a very thin crescent just before and after totality, its last few rays are twisted and contorted to produce the shadow bands. Although it has been assumed for years that shadow bands are caused by the Earth's atmosphere, this theory was only fully confirmed in the 1980s, by Johanan L. Codona of the

La Jolla Institute of the Center for Studies of Nonlinear Dynamics in California. Using computerized models of atmospheric turbulence, he was able to account for all their known characteristics.

Spotting shadow bands has eluded many veteran eclipse watchers, while others see them regularly. For instance, during the March 1970 eclipse that passed over Mexico and along the eastern coast of North America, many saw the bands rippling along the ground. Observing from Virginia, John E. Bortle of Stormville, New York, described the bands as 10 to 15 centimeters (4 or 5 inches) wide, spaced 30 centimeters (12 inches) apart, and moving at 6 to 9 meters (20 to 30 feet) per second from southwest to northeast.[1] Others reported that they were visible as much as four minutes before totality, and three minutes afterwards. Two and a half years later, while viewing the 1972 July 10 eclipse from 2,730 meters (9,000 feet) over Baker Lake (Northwest Territories, Canada), veteran eclipse chaser Jack Newton took a remarkable photograph of shadow bands dancing on the wing of a DC-3 aircraft (figure 3.3).

The greatest problem in viewing shadow bands is their low contrast against the ground. Many find shadow bands easier to spot against a white sheet laid flat on the ground. They are more difficult to spot if the sky is anything but crystal clear, though they have been seen under less than ideal sky conditions.

If you are successful in spotting the bands, be sure to announce your observation loud enough to be picked up by your tape recorder. Note the time that they were first seen, their direction of travel, and, if possible, their approximate speed. But do not become so consumed with looking for shadow bands that you miss other spectacular events such as the forthcoming Baily's Beads or the diamond ring!

The Shadow!

Just before second contact, look quickly toward the west for a glimpse of the onrushing shadow (figure 3.4). Note the time and direction that you first spot it, and try to describe the color and appearance on tape. Difficult to find the words, isn't it? Overhead, the sky turns an especially deep blue, while the horizon takes on an eerie yellow-amber-indigo-gray hue, almost like a 360-degree sunset.

Second Contact

As the Moon completely covers the Sun, its jagged edge breaks up the remaining razor-thin crescent of sunlight into a multitude of jewels, called *Baily's Beads*. Baily's Beads are named for Francis Baily, a British stockbroker-

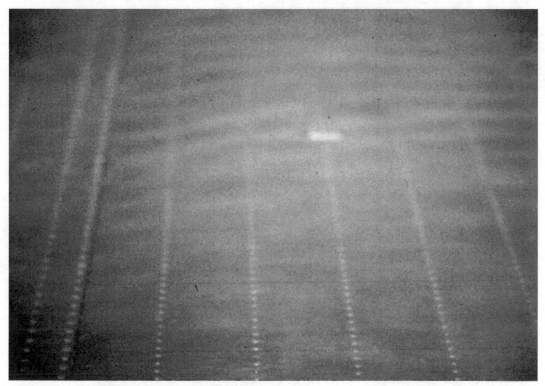

Figure 3.3 Shadow bands can be difficult to see and almost impossible to photograph. Flying over Baker Lake, Northwest territories, Canadian amateur Jack Newton photographed pronounced shadow bands on the wing of a DC-3 one minute after the end of totality during the 1972 July 10 eclipse. ($\frac{1}{125}$th second on High Speed Ektachrome [ISO 160].)

turned-astronomer who, upon seeing an annular eclipse in 1836, described the broken ring of sunlight as "a row of lucid points, like a string of bright beads." His report and description were published by the Royal Astronomical Society and immediately caught the attention of astronomers worldwide.

Each bead is the result of a point of light from the photosphere peering through a lunar valley, or between two of the Moon's mountain peaks. Their collective appearance depends on the varying terrain along the lunar limb, as well as the comparative sizes of the Moon and Sun. Shorter total eclipses always yield spectacular displays, an effect also enjoyed by observers situated near the edge of the path of totality.

As Baily's Beads vanish from view, cautiously remove your solar filter. The beads will flicker off one by one until a solitary beacon remains, looking like a brilliant flare burning on the rim of the Moon. This lone jewel, called the *diamond ring* (figure 3.5), appears as a dazzling celestial stone

Figure 3.4 The Moon's shadow retreating in the distance after bringing with it the excitement of the 1992 June 30 total solar eclipse to Punta Jose Ignacio, Uruguay. Photo by Bernard Volz. (16-mm fish-eye lens and Fujichrome 100 slide film.)

nestled against the jet-black silhouette of the eclipsed Moon. But look quickly, for the diamond ring lasts just a few seconds before it is extinguished as the Moon completely covers the Sun.

At the instant of second contact, use a telescope or high-power binoculars to look for the spiny chromosphere and any prominences that might be rising gracefully above the eclipsed limb of the Sun. Though the chromosphere is typically a deep ruby red, it can take on varying shades such as pink or even pale violet.

Totality

The emergence of one or more large prominences during totality always draws attention away from all else. With a harmony so perfect that it sounded rehearsed, everyone yelled "Look! Look at *that!*" during the July 1991 eclipse, as a huge prominence greeted thousands of eclipse-goers at the end of the second-contact diamond ring. Then, as if that weren't enough, another, perhaps even larger prominence, looking like a seahorse

Figure 3.5 The appearance of the Diamond Ring is one of the most dramatic events during a total solar eclipse. This photograph of the 1991 July 11 eclipse was taken by John Davis. (300-mm f/4 lens, 1/500th-second exposure on Kodachrome 64 slide film.)

in profile, slowly rose above the Moon's trailing (western) edge as totality progressed.

The Corona. A jubilant frenzy pierces the air. This is what everyone has been waiting for, what everyone has traveled so far to see: the sudden, stark beauty of a total solar eclipse.

First, examine the pearly white glow of the corona. Figure 3.6a shows the middle corona, while figure 3.6b portrays the fully blossomed outer corona. Astronomer José Joaquin de Ferrer was probably the first person to use the word *corona* (Latin for "crown") when describing the Sun's outer atmosphere.

The wide fields of view of binoculars are perfect for revealing faint coronal streamers extending one or more solar diameters beyond the eclipsed Sun. The narrower fields of view of telescopes often miss these tenuous filaments. Unless you have nerves of steel, it will probably be best to mount the binoculars on a tripod to steady the view.

a.

b.

Figure 3.6 The Sun's corona blossoms at the beginning of totality. Figure 3.6a, by Brian Kennedy, shows the middle corona (600-mm lens at f/8, ¼th-second exposure, Kodachrome 64 slide film), while Figure 3.6b by John Davis shows the full majesty of the outer corona (300-mm lens at f/4, ½-second exposure on Kodachrome 64 slide film).

Think of the solar corona as a celestial snowflake; each shows remarkable beauty, but no two are the same. If you are looking for a different kind of observing project that may also enhance your perception of fine eclipse detail, try sketching the corona. Even if your artistic skills do not put you in the same league as Picasso or Leonardo da Vinci, you just might get some pleasant results. Astro-artistry is surprisingly easy to do, and do well. And as almost any astro-artist will tell you, by drawing what you are observing, you will be able to detect subtle nuances that a casual observer might miss.

Begin your eclipse portrait by drawing a circle for the Moon's disk in the center of the paper. (You should, of course, do this well before the start of the eclipse!) Leave enough room for the full extent of the corona, which may stretch for two, three, or more solar diameters in all directions. Try to match its shape, structure, and extent as accurately as possible. Then, with the outline satisfactorily represented, study the corona's fine shadings and fluctuations, and translate these intricacies onto the paper. Remember, you are drawing a negative, so make the corona's brighter areas darker on the paper. You can also, of course, use black or dark paper and a white pencil, Conté stick, pastel, or crayon.

The Sky. As mentioned earlier, don't pass up the view of the whole sky, as the brighter stars and planets shine during this brief venture into midday twilight. Depending on their location in the sky relative to the Sun at the time of the eclipse, the five naked-eye planets are regular visitors during totality.

What of other visitors to the sky? On rare occasion, brighter artificial satellites, meteors, and even an errant comet might be visible during totality. Only four comets have even been documented during a total solar eclipse in modern times: Comet Tewfik, during the eclipse of 1882 May 17; Comet C/1947 F1 (Rondanina-Bester) during the eclipse of 1947 May 20; Comet C/1948 V1, seen at the 1948 November 1 eclipse; and, most recently, Comet C/1995 O1 (Hale-Bopp) during the 1997 March 9 total eclipse. The 1948 comet proved to be a fine naked-eye sight after it emerged from the Sun's glare a week later, and went on to be nicknamed "the Eclipse Comet of 1948."

The Environment. While a relatively small drop in temperature occurs during partial eclipses, total and annular eclipses can produce surprisingly large shifts. For instance, during the 1970 March 7 eclipse, the temperature at Norfolk, Virginia, fell from 15°C (59°F) at first contact to 4°C (39°F) at mid-totality. A slightly less dramatic change was felt from Baja California during the 1991 11 July total eclipse. Temperature at first contact was measured at 36°C (96°F). By the beginning of totality, the air temperature had cooled to 29°C (84°F).

The drop in temperature frequently leads to a change in the local meteorology. According to Jay Anderson, a meteorologist with Environment Canada who studies eclipse climatology, this may have an effect on local cloud cover: "It is never a good sign to have clouds in the sky before an eclipse, though some are worse than others. Good eclipse weather will be indicated by scattered cumulus clouds that often disappear after first contact. If there is a layer of heavier but mostly invisible moisture aloft, the dew point may be reached as the temperature drops, and extensive sheets of heavy cloud may form very quickly."

Panic.   This can be totality's most prominent feature, especially for first-timers. In a word, don't. Sure, totality passes quickly, but take it easy. If you feel anxiety about to strike, pause for a second, and take a deep breath. Jim Tomney writes of his harrowing experience during the March 1970 total eclipse: "In my haste to remove the solar filter so that I could view the eclipsed Sun, I accidentally bumped my telescope, causing the filter to become hopelessly wedged! I grabbed my lowest power and began sweeping the sky frantically for the Sun. Suddenly the target flew through the field, then back again. Try as I might, I just couldn't zero in on the eclipse because of the excitement."

Although the purpose of this chapter is to show just how much can be observed and accomplished during the few fleeting moments of totality, the message here is that *none of it is necessary!* Do as much or as little as you feel comfortable with.

Third Contact

Why is it that the last few minutes before quitting time on a Friday can seem to last all afternoon, but the few minutes of totality come and go in the wink of an eye? That is certainly what it feels like as, first, a brightening along the western edge of the eclipsed Sun appears, then bursts into the third-contact diamond ring, Baily's Beads, and the end of totality. Cheers go up as the main event ends and the thin crescent of the blindingly bright photosphere returns into view. The end of totality also signals the return of pre-totality viewing precautions, be it either a filter or projection.

Look quickly to the east as the receding lunar shadow races onward to darken another landscape for other observers. Then, with surprising speed, light and life return to the land as the Moon continues to slide silently off the Sun. Birds, having returned to their nests as the eclipse deepened, now awaken, undoubtedly confused about the shortest night they can recall. If you have been monitoring the changing climate, watch as the temperature slowly returns to near its pre-eclipse reading.

The partial phases after a total eclipse always seem anticlimactic. People begin to tear down their equipment even as the eclipse is still going on, all the while jabbering with great excitement about the dramatic event just witnessed. Personally, I never disassemble my telescopes or cameras until after the eclipse is officially over; doing otherwise just seems to be rushing nature needlessly. Follow through on any experiments, observations, and photographs begun during the partial phases that led up to totality. Enjoy the last few moments of the eclipse as the Moon finally releases the Sun at fourth contact.

ANNULAR SOLAR ECLIPSES

When the Moon appears smaller than the Sun during a central eclipse, it will leave a circle, or *annulus*, of sunlight even at maximum; this is known as an annular eclipse. The apparent size and thickness of the annulus depends on the circumstances. If, for instance, the Earth is near perihelion but the Moon is near apogee, then the annulus will look like a thick ring, as in figure 3.7. The closer the Moon and Sun appear in size, the thinner the annulus.

It is important to note that, regardless of the amount of Sun hidden during an annular eclipse, a solar filter or other form of eye protection must be used *at all times*.

The lead-in to an annular eclipse will seem much like that ushering in a total event. But when the crescent Sun is about half obscured, it becomes apparent that the eclipse will not be total, that the Moon will be too small to do the whole job.

Just at the instant of second contact (the beginning of annularity), look toward the trailing edge of the Moon's silhouetted disk as the lunar valleys and craters create a miniaturized display of Baily's Beads. Try to count the number of beads as they appear and disappear owing to the Moon's steady eastward progression. This same effect will also be visible along the Moon's leading edge at third contact.

During annularity, the Sun looks like a perfect ring if viewed from the center line, or a slightly lopsided bracelet if viewed closer to the northern or southern limits. While many people strive to be exactly on the center line of an annular eclipse to enjoy the event's perfect symmetry, spectacular views of a broken annulus are possible by viewing near the zone's edges.

Figure 3.7 The annular eclipse of 1994 May 10, as photographed by the author. (100-mm f/10 refractor and a neutral-density 5 solar filter, ⅟₆₀th-second exposure on Kodachrome 64 slide film.)

Patrick Poitevin of Belgium decided to do just that for the annular eclipse of 1995 April 29. "I chose to observe from Ecuador, but not from the center of annularity. Instead, I was along the path's edge. At maximum eclipse, a super view of the broken annulus, Baily's Beads, was visible. The ring on the lower part of the Sun was thicker, the upper part was very thin, small, and broken. The effect was just beautiful! I never understood why some 'crazy people' always went to the southern and northern limits of an [annular] eclipse. . . . Now I know."

Like totality, annularity can vary greatly in length, from as long as 12 minutes 30 seconds, to as short as a fraction of one second. The longest annular eclipse occurring between 2004 B.C. and A.D. 2526 did so on A.D. 150 December 7, when annularity lasted 12 minutes, 23 seconds. The longest annular eclipse listed in this book will occur on 2010 January 15 with an 11-minute, 8-second annular phase.

Extremely short central eclipses, where the Sun and Moon appear so close in size to one another that totality seems nearly imminent, are called *annular-total eclipses*. As explained in chapter 1, the Moon appears too small to mask the Sun entirely at the extreme ends of the eclipse's central path. Observers there will see a broken annulus of Baily's Beads, the result of the Moon's irregular surface (figure 3.8). Even though less than 1 percent of the photosphere may be shining through the distant lunar valleys, full safety precautions must be used to prevent eye damage. With great care, however, it should be possible to photograph the chromosphere and innermost corona, as Craig Small did in figure 3.9. But as the Moon continues in its orbit, the Earth's curvature causes its apparent size to enlarge just enough near the point of greatest eclipse to block all of the solar disk. Anyone nearby will witness a very short total eclipse.

The next annular-total eclipses will occur on 2004 October 14 and 2013 May 10. Because the paths of both eclipses are so narrow, you must make absolutely certain that you are on the center line; otherwise the central eclipse will pass you by!

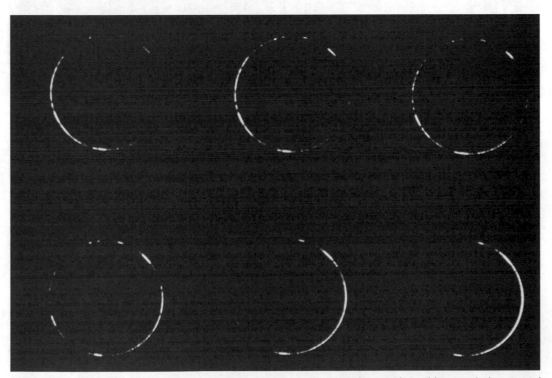

Figure 3.8 The 1984 May 30 annular eclipse was just about as close as it could get to being a total eclipse without actually being one! Notice how the multitude of Baily's Beads changed during the few seconds of annularity. This montage of six exposures was taken by Sam Storch. (Each ¹/₁₂₅th-second exposure, 500-mm lens and 2× teleconverter at f/16, Kodachrome 64 film, and a neutral-density 5 solar filter.)

Figure 3.9 The annular eclipse of May, 1984 was so close to a total eclipse that Craig Small was able to capture the chromosphere and innermost corona during the moment of maximum eclipse. (500-mm lens and 2× teleconverter at f/10 without a solar filter, 1/125th-second exposure on Kodachrome 25 film.)

THE "C" WORD

You've spent thousands of dollars and traveled thousands of miles, lugging expensive, delicate equipment around the world, only to discover that the solar eclipse you have so longed to see is being superseded by another eclipse—a total eclipse of the sky by clouds. Such is the fate of the eclipse chaser.

Just the mere suggestion of cloudy skies sends chills down an eclipse chaser's spine. What can be done? Some will jump into cars in hot pursuit of a hole in the clouds. Those on eclipse cruises hope that their ship's captain can maneuver the vessel into a clearing before totality strikes. Others go so far as to attempt to fly above the clouds in chartered airplanes.

No matter how well the weather maps are studied, no matter what contingency plans are put in place, sometimes the clouds just get you. Does this mean that the eclipse is a total loss? Not necessarily, as Richard

Sanderson of Springfield, Massachusetts, relates in his account of the 1977 October 12 eclipse.

"I traveled with a group of eclipse chasers to the village of Aguazul, in the foothills of the Andes in Colombia. Our group was situated near the end of the path of totality, where the Sun was fairly low in the sky. As luck would have it, even though much of the sky overhead was clear, a low-hanging cloud formation over the Andes completely obscured the Sun during totality. Still, the experience was memorable.

"About twenty minutes before totality, I noticed a small section of rainbow visible in some clouds to the southwest. The rainbow appeared normal until a few seconds before totality, when all colors faded except for a deep red hue. This red eclipse rainbow vanished with the onset of totality, and did not reappear again.

"With most of the western sky covered by scattered clouds, I was able to get a vivid sense of the umbra's movement as the distant clouds faded out one by one. Because of their varying distances, the clouds also gave the Moon's shadow a three-dimensional appearance; clouds within the shadow appeared as dark silhouettes against those located outside of the path of totality being illuminated by the Sun."

Miracles have been known to happen just before eclipse time. I was part of a team of nomadic astronomers who drove to North Carolina to view the annular eclipse of 1984 May 30. As we left home two days before the eclipse, it was pouring! In fact, it rained the entire way down, nearly 1,300 kilometers' (800 miles') worth. As we set up tents at our observing site, it continued to rain. We all awoke early on eclipse morning, partly from excitement, but mostly because of noise—it's hard to sleep in a tent being pelted with rain! By 8:00 A.M. the rain had stopped, but the cloud deck was solid, and prospects were gloomy. By 9:00 A.M. the clouds began to appear "layered," but unbroken. At 10:00 A.M., one hour before first contact, we could see one or two small breaks in the clouds (eclipse chasers call them "sucker holes"). Well, I don't know what happened in that next hour, but by first contact, at a little after 11:00 A.M., there wasn't a cloud in the sky, and we had a wonderful view of the entire event! Later we learned that just 50 miles to the east, the clouds never did clear.

But miracles don't always happen, as thousands of people who traveled to Hawaii for the 1991 July 11 total eclipse will attest. Though the weather was picture perfect the day before the eclipse, when it came time for the main event, much of the island was socked in with heavily overcast skies and, in some places, rain. Even with the clouds, some simple observations could be made, such as temperature and light drop, but those are little consolation for missing one of nature's grandest spectacles. Still, after the eclipse, one disappointed observer was heard to say, "I hate to lose an eclipse to clouds, but if I have to, let it be in Hawaii!" I applaud people like that, who can find happiness in the midst of such disappointment.

ECLIPSE ADDICTION

Yes, there is much more to the solar-eclipse experience than just the Moon passing in front of the Sun. It is the cooling of the air and the singing of birds. It is the appearance of stars in the sky, and the dark, silvery hues of the clouds. It is the illusion of sunset eerily encircling the entire sky.

For those who have never witnessed a total solar eclipse, the memory of your first minutes within the Moon's shadow will last a lifetime. But heed this warning: Once you see a total eclipse, you will await your next encounter with even greater anticipation. To Glenn Schneider, it's a craving that's not easy to kick: "No matter how much totality you've seen, it's never enough. Nicotine, alcohol, gambling, any conventional addiction you can think of; umbral dependence is worse."[2]

For many, the emotional impact catches them completely unprepared. Bob Buchheim of Coto de Caza, California, writes of the emotions that overcame him during the July 1991 eclipse, his first. "The power of the sudden onset of darkness, the precision with which the celestial bodies moved into position, the majestic impact of the scene overhead, was so overwhelming that I couldn't breathe without sobbing, and had to wipe tears from my eyes."[3]

4 A Bit of Luna-See

April 11, 1968, dawned like most other days. It was a Thursday, and I was looking forward to a long Easter weekend. I wanted to relax after what had been, at least through the eyes of a twelve-year-old, another trying week of school and homework. All this was about to change as I entered my sixth-grade science class that day. Against his usual policy, my teacher, Mr. Clark, had a weekend homework assignment posted on the blackboard. That Friday, he told us, there was going to be a total eclipse of the Moon. He wanted us all to watch it from our homes and write down our observations to hand in as a report on Monday. Just my luck, I thought.

I would have preferred to stay inside and watch one of the evening's television offerings, but instead I set up a card table and a lawn chair, and brought out a pair of small binoculars. The eclipse began right on time, just as Mr. Clark had promised. Slowly, over the next several hours, the Earth's shadow silently devoured the Moon, turning it a deep orange color around the middle of the eclipse. I don't remember the exact moment, but sometime during that eclipse, something clicked. I became so engrossed with the eclipse that I watched it in an almost hypnotic trance. That was it; I was hooked on the sky and just had to learn more.

Since then, I have grown into a lifelong lunar-eclipse lover. I already used the analogy of total solar eclipses and snowflakes: no two are ever the same. The same may be said of lunar eclipses. Each is unique in some way, thanks in large part to variations in the Earth's atmosphere and its influence on the appearance of our planet's umbral and penumbral shadows.

Coloration of the Earth's shadow, the result of sunlight refracting through our atmosphere, can change remarkably from one eclipse to another. At times the shadow is a brightly colored orange-red; at other times it appears a dark brown or gray. Still other eclipses have been recorded showing hints of purple, blue, or even green in the shadow. Other variations to the shadow include changes to its basic curved shape, such as unusual flattenings, notches, or even peaks.

Each lunar eclipse offers a unique visual experience. To better appreciate what may be expected during a lunar eclipse, let's review the various types of eclipses.

PENUMBRAL LUNAR ECLIPSES

Of all eclipses, both solar and lunar, penumbral lunar eclipses are the least spectacular. Indeed, the Earth's outermost shadow is so weak that a penumbral eclipse may come and go completely unnoticed by casual observers.

All this is not to say that penumbral eclipses are boring. On the contrary, their delicate nature will test an observer's skills and acuity. Several noteworthy observations can be made as the Moon slips deeper into the penumbra.

As with all eclipses, we should expect some penumbral events to be more impressive than others. *Shallow* penumbral eclipses, broadly defined here as having a penumbral magnitude of less than 0.7 (that is, half of the moon's diameter within the penumbra), are difficult to detect, and attract little attention from the astronomical community. *Deep* penumbral eclipses are referred to in this context as having penumbral magnitudes greater than 0.7, and are typically visible with or without optical aid. These latter events give observers the greatest opportunity.

Though the task may sound simple, marking the time when the penumbra is first detectable proves to be one of the greatest challenges for observers. Many observers find it easier to detect the penumbra by viewing the dimmed Moon with a Moon filter (a filter used to dim the Moon's dazzling appearance) attached to their telescope's eyepiece. Others find it easier to detect the penumbra by sweeping their telescopes back and forth across the Moon's surface at moderate power (say, 150×).

Also record in which direction on the lunar limb the penumbra was first seen. Photocopy the map of the Full Moon found in figure 4.2. On the photocopy, draw the location of the penumbra's edge at your first sighting. Continue to note the progress of the penumbra on the same map, drawing in its leading edge as the eclipse continues. Be sure to note the time of each.

Does the penumbra show any color? Although many of us see it as simply a light gray, the penumbra can actually exhibit many extremely subtle color variations to keen-eyed observers, including tannish yellow, gray, sandy, tan-gray, yellow-brown, brownish gray, and bluish gray.

Note the influence the penumbra has on the appearance of surface features, especially the brighter areas, such as craters' ray systems. Tony Royal, observing the deep penumbral eclipse of 1987 October 6–7 from Port Saint Lucie, Florida, paid close attention to the weak shadow's effect on the crater Tycho and its vast ray system. His report in *Sky & Telescope* magazine noted that "while I didn't see any change in the crater itself, the rays lost much of their brilliance as the eclipse progressed."[1]

It is interesting to note that, during a penumbral eclipse, an astronaut on the Moon's surface would see a partial solar eclipse, with the Earth blocking a portion of the Sun's disk. During partial and total lunar eclipses, our astronaut friend would witness total solar eclipses, with the Earth masking the entire Sun, including the corona, at maximum. In each case, since the Moon is gravitationally locked in its orbit (that is, the same side of the Moon always faces the Earth), the Earth would neither rise nor set, but instead remain perfectly still as the Sun passed behind it. (A note to nitpickers: the Earth does move slightly in the Moon's sky, owing to an effect called *libration*, which causes the Moon to wobble slightly on its axis. Here on the Earth, the Moon's librations allow astronomers to view slightly more than half of the lunar disk.)

PARTIAL LUNAR ECLIPSES

Partial eclipses of the Moon, even those when the Moon will not be covered fully by the umbra, are spectacular events that always draw wide attention among amateur astronomers. Frequently, members of astronomy clubs will gather together for an "eclipse party" to share the excitement of watching the Moon silently slipping partway into the Earth's shadow.

There are many things to do and look for during a partial lunar eclipse (figure 4.1). Besides photography (dealt with separately in the next chapter), observers can time the exact moments when the umbra first touches and last leaves the lunar disk, monitor the changing colors and any unusual irregularities in the umbra, and time when the umbra crosses over certain select lunar features.

Timings

Trying to catch the exact moment of first contact, when the umbra first touches the Moon's leading (eastern) limb, is much more challenging than

Figure 4.1 The show begins! Photo by Richard Sanderson. (150-mm f/12 refractor, ⅟₃₀th-second exposure on Kodachrome 64 slide film.)

it sounds. You might think the experts should already know that, since the beginning and ending of an eclipse can be determined far in advance. But science can only go so far. While predictions for eclipses can be derived far in advance, they are based solely on mathematics. As far back as 1707, astronomers discovered that the Earth's shadow appears larger than could be explained by simple geometry, and that its size changed from one eclipse to the next. Studies show that high-altitude clouds and moisture, ozone, volcanic dust, our atmosphere's *oblateness* (the fact that the atmosphere is not perfectly round, but rather bulges at the Equator), and even leftover debris from meteors being incinerated in our upper atmosphere, all influence the shape and size of the umbra. The only way to learn more about this mystery is through systematic timings of shadow contacts and of the shadow's passage over certain key lunar-surface features.

For the timings to be of scientific value, they must be accurate to at least 0.1 minute (6 seconds). A modest-sounding task, but one that often proves daunting because it can be very difficult to judge exactly where the penumbra ends and the umbra begins! That's the key. To help determine this often vague boundary, veteran eclipse watchers offer several hints, in addition to those mentioned earlier.

Frequently, small-aperture telescopes can yield better timing results, because the greater light-gathering power of larger instruments (those greater than 8 inches [20 cm] in diameter) makes the Full Moon overwhelmingly bright, muddling the distinction between the umbra and penumbra. Regardless of telescope aperture, use your lowest power so that the entire lunar disk fits into the field of view, with room to spare.

Higher magnifications can also blur the distinction between the umbra and penumbra. To enhance shadow contrast further, consider using a neutral-density Moon filter.

Next, you will need to know the *exact* time in order to judge the times of contact. Going by regular broadcast radio stations, even if they use a seemingly accurate chime on the hour and half hour, just isn't good enough. Instead, the National Bureau of Standards broadcasts precise time signals from their offices in Boulder, Colorado, on shortwave radio station WWV (and rebroadcast over WWVH from Hawaii). Canada offers a similar service over its shortwave radio station CHU. Further details on these and other ways of obtaining precise time signals are offered in the discussion on occultation timing, found at the end of this chapter.

As with contact timings, crater timings are best done with small apertures, neutral-density Moon filters, and magnifications of 100× or less. The object here is to mark the shadow's passage over the center of the crater to within an accuracy of at least six seconds. Because the shadow moves across the Moon so quickly, smaller craters are easily timed; larger craters, on the other hand, are more challenging. For these, mark the moment when the shadow first touches the rim, and again when the shadow touches the opposite side of the rim. Afterwards, average the timings to get the moment when the shadow crossed the crater's center.

For those who want to try their luck with crater timings, figure 4.2a shows twenty of the Moon's most prominent craters, as suggested by the Association of Lunar and Planetary Observers (ALPO). Compare the map to the photograph of the Full Moon in figure 4.2b.

Begin your preparation for crater timing several Full Moons in advance by getting to know the surface of our satellite. Using the same equipment as for the eclipse, locate and identify each of the features on the map. Learn to recognize them first from the map, then only by sight. That way, each crater will become an old friend, so that by the time the eclipse begins, you won't miss one during the excitement of the moment.

As mentioned earlier, many observers prefer to record their observations on a reliable tape recorder for later analysis. Position the microphone so that it will pick up both the shortwave radio's time signals and your

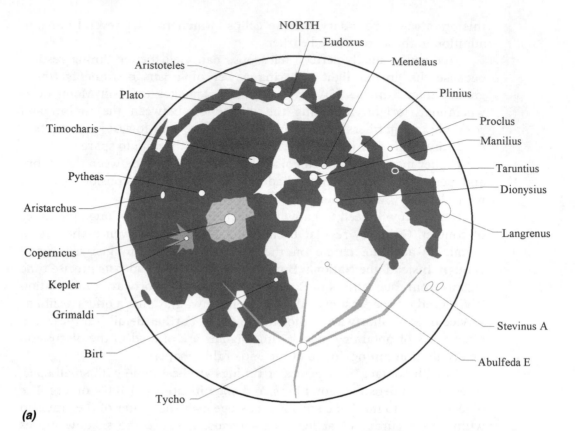

NORTH

Eudoxus
Aristoteles
Menelaus
Plato
Plinius
Timocharis
Proclus
Manilius
Pytheas
Taruntius
Aristarchus
Dionysius
Copernicus
Langrenus
Kepler
Grimaldi
Stevinus A
Birt
Abulfeda E
Tycho

(a)

Figure 4.2 (a) Map of twenty of the Moon's most prominent features, often used for crater timings. Compare the view with (b) the photograph of the Full Moon taken by Richard Sanderson. First, identify each feature on the photograph, then find them on the Moon itself using a telescope.

voice, as described in the discussion of grazing occultations, later in this chapter. Then, as the eclipse progresses, announce when the shadow contacts each of the craters in your program.

An increasingly popular alternative method is to videotape the eclipse through a telescope, using a small video camera; others prefer a lightweight camcorder for later analysis (figure 4.3). Position the shortwave radio near the camera's microphone so that the time signals will be recorded on the audio track as the Moon's disappearing act is captured on video.

Byron Soulsby of Calwell, Australia, has spent years studying the phenomenon that causes the umbra to be larger than calculated mathematically. Since the total lunar eclipse of 1972 January 30 Soulsby has analyzed 27 lunar eclipses, including many not observable in Australia. His detective work into the mystery of the too-large umbra has shown that the Earth's atmosphere may be more oblate than originally thought. Although earlier re-

Figure 4.2(b)

searchers concluded that the Earth's umbra was out of round by $\frac{1}{139}$ (that is, the change in height of the Earth's atmosphere at the equator as compared to the poles), Soulsby's findings indicate an oblateness of $\frac{1}{102}$. Small numbers, yes, but it would mean that our atmosphere is nearly three times more oblate than the "solid" Earth.

To fuel his continuing research, Soulsby regularly receives observation reports from amateur and professional astronomers around the world. If you would like to learn more about how you can contribute to his research, write to Soulsby at Calwell Lunar Observatory, 23 Andrew Crescent, Calwell, ACT 2905, Australia.

Appearance

Take a look at the shape of the shadow's leading edge as the umbra slides across the lunar surface. Does it look uniformly curved, or blunted? The latter strange effect is known as *cusp extensions* (figure 4.4). Cusp exten-

Figure 4.3 A small video camera can be used to record the passage of the Earth's shadow across the Moon for later analysis. Photo by Byron Soulsby, Calwell Lunar Observatory.

sions are bright elongations, or "horns," created by the Earth's shadow giving the appearance of being flattened over the central portion of the lunar disk and curving off suddenly near the lunar limb. One cusp extension will frequently appear more prominent than the other.

Another feature to make note of is the shadow's color. Often observers will look at the shadow during the partial phases and pronounce it as either dark brown, dark gray, or black. This is what you will probably see unless the shadow is isolated from the portion of the Moon still uncovered by the umbra. Isolating your telescope's field of view only on the eclipsed portion of the Moon will enhance subtle colors and tints. The leading and trailing edges of the umbra are frequently characterized as coppery red, although some eclipses have worn a distinct blue or green fringe.

TOTAL LUNAR ECLIPSES

Total lunar eclipses present unique observational opportunities for everyone. Many people make systematic observations of the Moon's appearance

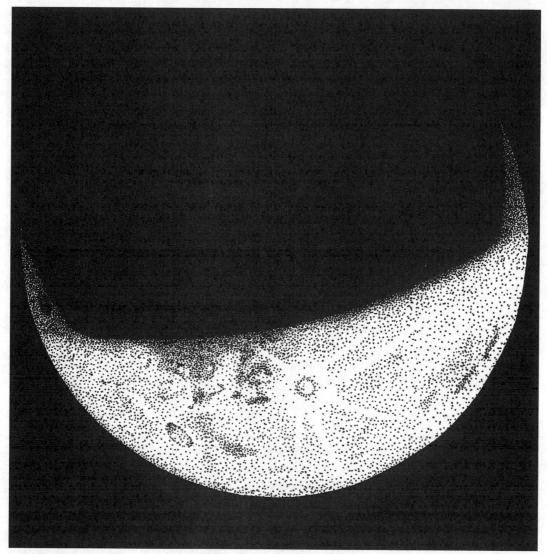

Figure 4.4 Russ Sampson of Edmonton, Alberta, noted an unusual appearance to the umbra's edge during the 1992 June 14–15 partial eclipse. His drawing, made with a 75-mm (3-inch) refractor at 20×, shows bright cusp extensions into an otherwise pitch-black umbra.

during totality by rating both its apparent magnitude and its luminosity (see instructions below). Others carefully monitor the Moon's progress through the umbra by timing the Moon's passage over distant stars (see "Timing Occultations," later in this chapter). Some use the short period of darkness to search for new comets or make observations of existing comets and variable stars that are otherwise lost in the Full Moon's glare. (Most recently, the total lunar eclipse of 1996 April 4 gave sky watchers a chance to view the brilliant Comet Hyakutake in a dark sky.) There are those who enjoy capturing the moment on film (detailed in the next chapter) or on canvas, while many others just sit back and enjoy the show.

Appearance

Many variables, such as the Moon's location compared to the center of Earth's umbra and the clarity of the Earth's atmosphere, contribute to make each lunar eclipse a unique and memorable event. But how can they best be summarized and analyzed, and the data used to compare one eclipse to another?

One of the simplest and most telling total lunar eclipse observations is to estimate the Moon's luminosity value at various times during totality. Early in the twentieth century, the French astronomer André-Louis Danjon devised a clever five-point scale for rating the darkness of a total lunar eclipse. The Danjon scale, reproduced in table 4.1, has since gone on to become the standard by which all total eclipses are judged.

The Danjon scale should be used to estimate the appearance of a total eclipse *only* with the naked eye. Surprisingly, the often subtle colors of a total eclipse tend to fade as magnification increases. The late Joseph Ashbrook suggested that this might be due to a reduction in surface brightness to near or below the threshold for color perception.[2]

Although Danjon devised the scale to fit every lunar eclipse, it is rare for an eclipse to match one of his descriptions exactly. Instead, most seem to fall somewhere between two values—and therefore so should your estimate. For example, at mid-eclipse, if the Moon appears a muddy clay color highlighted with a bright, almost yellow rim, then the Danjon luminosity value would fall somewhere between 2.0 and 3.0. After carefully examining the Moon's appearance, judge which description it most resembles, then

Table 4.1 **Danjon Lunar Eclipse Luminosity Scale**

$L = 0.0$	Very dark eclipse. Moon almost invisible, especially at mid-totality.
$L = 1.0$	Dark eclipse, gray or brownish coloration; lunar surface details distinguishable only with difficulty.
$L = 2.0$	Deep red or rust-colored eclipse; central part of the umbra dark, but outer rim of the umbra relatively bright.
$L = 3.0$	Brick-red eclipse, usually with a brighter (frequently yellow) rim to the umbra.
$L = 4.0$	Very bright copper-red or orange eclipse, with a bluish, very bright umbral rim.

prorate the value accordingly. In this case a luminosity value of 2.4 or 2.5 would seem most appropriate. Astronomical magazines always encourage readers to send in their observations and reports, but to be of real value, always include your time, location, instrument used, if any, and a description of the sky conditions.

A less-often-quoted rating system was introduced in 1924 by Willard J. Fischer. The Fischer scale uses a three-point strategy based not on color but on the amount of surface detail visible through various instruments. Table 4.2 offers a summary.

Another interesting total-lunar-eclipse project is to estimate the change in the Moon's apparent magnitude. Like all extended (i.e., nonstellar) objects, the Moon's apparent magnitude is based on how bright it would appear if its half-degree disk were somehow condensed to a starlike point. Normally the Full Moon shines at magnitude −12.7. How faint does it get during a total eclipse? Any guesses? −4? −2? 0? Actually it can get much dimmer than that, but trying to estimate the exact brightness of the Moon is not as easy as it may sound.

First you have to shrink the Moon to a point of light. Observers have come up with a number of different and clever ways to do this, such as observing the reflection of the Moon in a convex mirror, a chromed "baby moon" hubcap, or even a silver Christmas-tree ornament. Though all of these objects work, I've had the greatest success looking through a pair of binoculars . . . backwards. The Moon will shrink by the inverse of the binoculars' magnification (e.g., through 7× binoculars, the Moon will appear one-seventh as large as with the naked eye, or approximately 4 arcminutes across).

Aim the binoculars toward the Moon and look through one of the objective lenses with one eye. With the other (unaided) eye, look around the sky for suitably bright stars to judge the Moon against. Best results are had by comparing the Moon against two stars, one a little brighter and one a

Table 4.2 Fischer Lunar Eclipse Scale

Grade	Description
2	The naked eye sees "spots" on the eclipsed Moon, and the seas and other detail can be seen with hand instruments (small binoculars and field glasses).
1	Instruments of 5 cm (2 inches) up to 15 cm (6 inches) are necessary to show detail on the eclipsed surface.
0	Apertures greater than 15 cm (6 inches) are needed to see surface detail.

little fainter. If possible, use comparison stars that lie at more or less the same altitude above the horizon as the Moon, to ensure that atmospheric interference does not unduly skew the estimate. Find the stars' apparent magnitude (some potential comparison planets and stars, broken down by northern-hemisphere season, are listed in table 4.3), judge which the Moon appears closest to, and prorate your magnitude estimate accordingly.

Comparing a binocular-reduced Moon to a star in the naked-eye sky cannot be used in a direct, one-to-one correlation; instead, a correction factor must be introduced to compensate for the dimming effect caused by the reversed binoculars. This correction factor is then subtracted by the perceived magnitude of the Moon through the binoculars to arrive at the Moon's true apparent magnitude. Here is how it works.

The most difficult part is coming up with the correction factor (F) for your pair of binoculars. Table 4.4 lists the correction factors for several of the more popular binocular magnifications; these are only approximations, however. You should try to come up with your own correction factor that fits your particular binoculars.[3] To check the accuracy of your correction factor, compare the view of, say, Venus, Jupiter, or Sirius through the reversed binoculars with naked-eye comparison stars.

If all this is still a bit befuddling, perhaps a real-life observation will help clear things up. For the 1993 November 28 total eclipse, I attempted to estimate the apparent magnitude by glancing back and forth between the Moon, shrunken through a pair of 7× binoculars, and stars in the naked-eye sky. After comparing stars of known magnitudes with the miniaturized Moon, I found the lunar brightness to be approximately equal to the 3.0-magnitude star Sadalmelik in Aquarius. To get the Moon's apparent magnitude, I still had to use the correction factor to compensate for the effect of the binocular shrinkage. Using a correction factor of 4.5, I estimated the Moon's actual apparent magnitude (M) to be

$$M = m - F$$
$$= 3.0 - 4.5$$
$$= -1.5$$

where M is the corrected apparent magnitude and m is the magnitude estimate through the binoculars.

This method works well provided you can see suitable comparison stars, but what if the Moon is so dim that it literally disappears from view through the binoculars? In cases such as these, use a telescope or binoculars (looking through them the right way!) to compare the Moon with suitable comparison stars following a technique used by comet observers called the "Sidgwick Method." This was first presented in the book *Obser-*

Table 4.3 **Comparison Planets and Stars**

Name	Star Designation	Apparent Magnitude
Venus	—	−4*
Jupiter	—	−2*
Spring		
Arcturus	α Boötis	−0.1
Spica	α Virginis	+0.9
Regulus	α Leonis	+1.4
Denebola	β Leonis	+2.1
Zosma	δ Leonis	+2.6
Altarf	β Cancri	+3.5
Summer		
Vega	α Lyrae	0.0
Antares	α Scorpii	+0.9
Shaula	λ Scorpii	+1.6
Rasalhague	α Ophiuchi	+2.1
Ascella	ζ Sagittarii	+2.6
—	ι Scorpii	+3.0
Autumn		
Fomalhaut	α Piscis Austrini	+1.2
Hamal	α Arietis	+2.0
Sharatan	β Arietis	+2.6
Sadalmelik	α Aquarii	+3.0
Homam	ζ Pegasi	+3.4
Winter		
Sirius	α Canis Majoris	−1.4
Rigel	β Orionis	+0.1
Betelgeuse	α Orionis	+0.4
Pollux	β Geminorum	+1.1
Castor	α Geminorum	+1.6
Saiph	κ Orionis	+2.1
—	ζ Tauri	+3.0
Wasat	δ Geminorum	+3.5

*Because the magnitude values for Venus and Jupiter can vary noticeably from these values, readers are urged to consult the current issue of one of the astronomical periodicals or annuals listed in Appendix B for the precise magnitude value at the time of the lunar eclipse being studied.

Table 4.4 **Reversed-Binocular Magnitude Correction Factors**

Binocular Magnification	Correction Factor
6×	$F = 4.2$
7×	$F = 4.5$
8×	$F = 4.8$
10×	$F = 5.3$
11×	$F = 5.5$
12×	$F = 5.7$
16×	$F = 6.3$
20×	$F = 6.8$
25×	$F = 7.3$
30×	$F = 7.7$

vational Astronomy for Amateurs (Enslow, 1982) by J. B. Sidgwick. This method requires the observer first to memorize the brightness of the eclipsed Moon through binoculars or a telescope, then to swing the instrument to a star of known brightness. The star's image is then quickly defocused until its size matches that of the in-focus Moon. Now compare, in your mind, the brightness of the out-of-focus stars with the memorized brightness of the eclipsed Moon. Does the Moon appear brighter or fainter? Continue going back and forth between the in-focus eclipsed Moon and out-of-focus comparison stars until a reliable magnitude value for the Moon is obtained.

Table 4.5, compiled from published reports[4] as well as from personal notes, shows how some of the eclipses over the past several decades have rated on both the Danjon Luminosity Scale as well as in apparent magnitude.

Though it is useful for deriving scientific findings, reducing a total lunar eclipse to just a set of numbers is really an injustice to an event that is full of unexpected sights and surprises. One of the darkest total eclipses in recent memory was that of 1992 December 9–10. With the Earth's shadow influenced by the volcanic aerosols from the June 1991 eruption of Mount Pinatubo in the Philippines, the Moon nearly disappeared from view during totality. Observing from outside New York City, Joe Rao notes that this eclipse had "the greatest brightness variation" of the many that he has witnessed. "I have never seen the Moon go through so many contortions of light and color. It was a most remarkable event."

Twenty-nine years earlier, the 1963 December 30 eclipse set a new standard for dark eclipses. Thanks to volcanic aerosols released during the eruption of Mount Agung in Bali, the Moon completely disappeared during

Table 4.5 **Selected Eclipse Performances: 1960 through 1996**

Eclipse Date	Umbral Magnitude*	Danjon Value†	Apparent Magnitude‡
1996 September 27	1.24	2.5	−3.0 to −2.0
1996 April 4	1.38	2.5	−2.0 to −1.0
1993 November 28–29	1.09	1.8	−0.5 to −1.5
1992 December 9–10	1.27	0.2	+3.0 to +3.5
1989 August 16–17	1.60	2.0	−2.0 to −1.0
1986 April 24	1.21	3.0	−3.0 to −2.0
1985 May 4	1.24	1.7	−0.5 to +0.5
1982 December 30	1.19	0.3	+3.0 to +3.5
1982 July 6	1.72	1.5	+1.5 to +2.5
1982 January 9	1.34	2.8	−3.5 to −2.5
1979 September 6	1.10	3.0	−2.0 to −1.0
1978 September 16	1.33	2.2	−1.5 to −0.5
1978 March 24	1.46	2.1	−2.0 to −1.0
1975 November 18–19	1.07	2.6	−3.0 to −2.2
1975 May 24–25	1.43	0.7	+0.5 to +1.0
1974 November 29	1.29	2.5	−2.0 to −1.0
1972 January 30	1.06	2.9	−3.0 to −2.5
1971 August 6	1.73	2.0	0.0 to +0.5
1971 February 9–10	1.31	2.6	−2.0 to −1.5
1968 October 6	1.17	1.6	0.0 to −1.0
1968 April 12–13	1.12	2.3	−2.0 to −1.0
1967 October 18	1.15	3.2	−1.5 to −0.5
1964 December 19	1.18	1.6	0.0 to +0.5
1963 December 30	1.34	0.0 to 0.3	+4.0 to +4.5
1960 September 5	1.43	1.8	−0.5 to 0.0
1960 March 13	1.52	1.9	−1.0 to −0.5

Notes:

*Umbral magnitude of an eclipse is a measure of how deeply immersed the Moon is in the Earth's umbra, and does not refer to its apparent brightness.

†Danjon values are mid-eclipse estimates, when the Moon is typically its darkest. Figures listed in the table have been compiled from various sources, including published reports and personal observations.

‡Apparent magnitude *is* a measure of how bright an object appears in our sky—in this case, the Moon at mid-totality.

totality when viewed with the naked eye. Even those peering through binoculars and telescopes had difficulty seeing the lunar disk at times. Nearly all observers commented on the blackness of the Earth's shadow that night, though a few also reported somber shades of brown, blue, and red on the leading and trailing edges of the umbra. Many compared the

Moon's overall naked-eye brightness to that of M44, the Beehive open cluster in Cancer the Crab (M44 shines at apparent magnitude 3.1).

The brightness of a total eclipse is also dependent on how near the Moon is the center of the Earth's umbra. Presumably, the umbra's central area is less polluted with refracted sunlight, leading to darker eclipses; brighter eclipses are more apt to occur when the Moon passes near the umbra's edge. The total eclipse of 1996 April 3–4 (figure 4.5) is a good example, when the Moon passed south of the umbra's center to produce a bright rusty orange Moon.

Other unusual anomalies are often found lurking in the shadow of the Earth. For instance, though not as dark as some more-recent eclipses, the dusky total eclipse of 1960 September 5 was highlighted by a peculiar dark band seen to shift across the Moon's face as the eclipse progressed. As with most shadow irregularities, these unusual features were likely caused by clouds or dust in the Earth's atmosphere along the sunrise-sunset line.

Timing Occultations

In addition to trying to capture the character of an eclipse both in words and in numbers, one of the most worthwhile observation programs that an

Figure 4.5 The 1996 April 4 total lunar eclipse was one of the brightest eclipses in decades. Photo by Paul Whitmarsh. (200-mm f/10 Schmidt-Cassegrain telescope, 30-second exposure on Fujichrome Provia ISO 400 slide film.)

Figure 4.6 Lunar eclipses give observers a chance to time occultations that would otherwise be invisible because of the Full Moon's overwhelming brightness. Photo by Richard Sanderson. (300-mm lens at f/8, 4-minute exposure on Ektachrome P800/1600; camera was piggybacked on a clock-driven mounting).

amateur can engage in is timing the occurrence of *lunar occultations*. These occur when the Moon passes in front of a star (figure 4.6). Normally, around the time of Full Moon, you can't see the Moon pass over a star because the Moon is so bright. During an eclipse you may watch as the Moon comes closer and closer to the distant point of light until, just as it is touched by the advancing lunar limb, the star winks out. A star passing centrally behind the Moon will remain hidden for a little over an hour. The farther north or south one is from this central region, the shorter the occultation's duration. Too far north or south, and the Moon will miss the star entirely.

A captivating sight is visible along the northern and southern edges of an occultation's region of visibility across the Earth. There the Moon will appear to graze over the star, alternately covering and uncovering the point of light as the star passes behind lunar mountains and valleys. Depending on the ruggedness of lunar topography along the limb, such a *grazing occultation* may only be visible only from a narrow zone a few kilometers wide.

Grazing occultations are best observed by a team of several observers with telescopes and timing equipment set up at specified positions perpendicular to the occultation's path. Each team member times the star's disappearing act from his or her location. Later, by knowing the observers' exact positions relative to one another, the team's data can be used to determine the details of the lunar profile as well as refine the position of the Moon relative to the star.

Why is the timing of occultations important? David Dunham, founder of the International Occultation Timing Association (IOTA), writes: "Occultations are useful for refining knowledge of the positions and motions of stars, and can be used to improve parameters such as the tilt of the Earth's equator relative to the ecliptic. Advancing our knowledge of the lunar profile aids in the analysis of total solar eclipse timings, which can then be used to study small but climatically significant changes in the diameter of the Sun over periods of many years. Also, a star's disappearance or reappearance may occur in steps, indicating a previously undiscovered close double star that cannot be detected by direct observations." Indeed, this is how the brilliant stellar jewel Antares in Scorpius was discovered to be a double star.

Critical to the usefulness of occultation timing is knowing the precise location of the observer to an accuracy of about 10 feet (3 meters) relative to landmarks (such as road intersections or large buildings) that are shown on a detailed topographic map of the area. From these a precise latitude and longitude of each observation site can be measured from the map. The United States Geological Survey (USGS) has produced an impressive set of maps of the entire country, enabling an observer to determine his or her exact geographical location in any of the fifty states. For ordering information, see Appendix I. Another "high-tech" way of determining your exact position is to use a Global Positioning Satellite (GPS) receiver, available at most boating and marine supply stores.

Timing occultations requires quick reaction time as well as a telescope, a stopwatch, and a shortwave radio to pick up precision time signals from either WWV or WWVH (broadcast on 2.5, 5, 10, 15, and 20 MHz), or CHU (heard on 3.330, 7.335, and 14.670 MHz). WWV time signals are also available from the National Bureau of Standards in Boulder, Colorado, by telephoning (303)499–7111.

Dunham notes, however, that all team members observing an occultation do not need a portable shortwave radio. Instead, as the occultation progresses, one person can make a master tape recording of WWV and, simultaneously, a local broadcast of an AM or FM radio station. Other observers in the team can then use and record the same broadcast radio station for their own occultation "hits and misses."

With the observing station set up, peer through your telescope at the Moon and the soon-to-be-occulted star. As the time of the occultation draws near, start the stopwatch as an exact minute mark is announced by the time-signal radio station. The instant the star disappears, stop the watch and check the accrued time. To find the exact moment when the star disappeared, add that figure to the time you started the watch. Do the same for the star's reappearance.

You will find that timing reappearances is much more difficult. Unless you have obtained exact data for your observing site, you can only guess when and where the star will pop back out. Fortunately, IOTA supplies precise occultation predictions to its members. For information on joining the association, write to the address listed in Appendix C.

Grazing occultations, as well as multiple occultations occurring over a short window of time, require a different approach. For this you will need a shortwave radio and a reliable tape recorder. Set the recorder and the radio near each other, adjusting the volumes until both the radio and your voice are distinguishable on the tape. Then, at the instant the star disappears, shout "D" (for disappearance) loud enough for your tape recorder to pick up over the radio. As soon as the star reappears, shout "R," and so on for each event. (Some observers prefer to use either "off" and "on" or "out" and "in" to signify a star's disappearance and reappearance; others use beepers or a variety of electronic or mechanical noisemaking devices. The choice is yours, as long as the tone is sharp and quick.) Visual timings made to an accuracy of half a second are quite adequate for defining the lunar profile; for a graze, knowing the observer's location is more sensitive than the timings, since observers even 15 meters (50 feet) apart will notice differences in their event timings.

Astronomical periodicals, such as *Sky & Telescope* and *Astronomy*, publish information on occultations typically a month or two before an eclipse is set to occur. Additional information is available from IOTA, as well as the International Lunar Occultation Center (ILOC), Geodesy and Geophysics Division in Tokyo. Updates on any late-breaking changes in an occultation's predictions can be heard by calling IOTA's special telephone "hot line" number. (See Appendix C for addresses and telephone numbers.)

Lunar eclipses are exciting events to witness both from an aesthetic perspective as well as from a scientific point of view. By carefully planning an observing program long before the main event, you can gather valuable data during the Moon's progress through the Earth's shadow. Or you can simply sit back and enjoy the event's beauty. The choice is yours.

5 Eclipse Photography

Very few people view an eclipse without trying their luck at photography. Happily, unlike many other celestial objects and events, eclipses offer something for just about every photographer. Depending on what aspects you want to record for posterity, you probably already own a camera that will work well. Of course, it takes more than just a camera to photograph an eclipse; tripods, cable releases, and a variety of lenses are just the beginning.

EQUIPMENT AND TECHNIQUES

Still Cameras

For most eclipse photographers, 35-mm single-lens reflex (SLR) cameras are the best choice. They offer a greater variety of auxiliary lenses and attachments than any other style of camera.

Today's typical SLR is a very sophisticated piece of equipment, with automatic exposure, focus, film advance, flash, and so on. But such fancy cameras are *not* required to photograph an eclipse; indeed, they can be a hindrance. To capture the full complement of eclipse phenomena on film, you'll need a camera that is *manually* adjustable—that is, one in which both shutter speed and lens aperture can be set by the photographer. While completely automated cameras will be able to capture some aspects of an eclipse, their onboard computerized "brains" were designed more for photographing Aunt Martha sitting by the family garden than for shooting the Sun or Moon!

Other photographers swear by larger-format cameras (those that use larger film formats than 35 mm, such as 120 roll film or even film plates). Admittedly, the larger film does permit greater resolution compared to film grain size, but their higher prices and limited film availability usually make these cameras less attractive to most casual photographers.

Regardless of format, if you are running on a limited budget and want to buy a camera and lens to photograph an eclipse, sink as little money as possible into the camera—but invest heavily in the lens. A 20-year-old used camera combined with a high-quality lens will take better photos than a brand-new camera with a cheaply made lens.

Lenses

Many books on astrophotography would have you believe that because of their small size in our sky, the Sun and Moon are impossible to photograph without a long focal-length lens. That shortsighted advice is simply not true. All lenses, regardless of focal length, are capable of capturing unique aspects of an eclipse, but it is important to understand their pros and cons. The secret to successful eclipse photography is knowing how to play these to your advantage.

The first step is knowing just how *small* the image of the Sun or Moon will appear on the film. Remember, from our earthly vantage point, each subtends only half a degree across—about the same size as a penny (about 19 centimeters, or 0.75 inch in diameter) seen at a distance of about two meters (a little more than six feet). That's tough enough to see, let alone photograph! As a rule of thumb, you can estimate the actual size of the solar or lunar image on film by dividing the lens's focal length by 109. For example, a 100-mm lens will produce an image that is just under 1 mm across—about the thickness of a United States penny. Small? Yes, but not unworkable. If taken with a high-quality lens, the image can be enlarged enough to show the Moon's ruddy disk nestled among a star-filled field or the long, filamentary fingers of the solar corona.

Seeing is believing, so take a look at figure 5.1. This full-frame montage shows the Moon as photographed four times with the same 35-mm camera. The only difference is in the focal lengths of the lenses used. Gauge the size of the image with each lens to determine which is most suitable for the aspect of the eclipse that you want to capture. (Since the Moon and Sun appear effectively the same size in our sky, this illustration may also be used to judge the proper focal length for solar eclipses. But total solar eclipse photographers, take note: if you are interested in capturing the full glory of the Sun's outer corona, estimate its diameter to be at least twice that of the eclipsed solar disk.)

Figure 5.1 Just how large will the Moon or Sun appear in a photograph? This Full Moon montage will give you some idea. The smallest image was photographed with a 50-mm focal-length lens, followed by 100-mm, 400-mm, and 1000-mm focal lengths.

The proper choice of lens depends on what facet of the eclipse interests you, but keep in mind that every focal length, from wide field to telephoto, is useful. As an aid in the decision-making process, the following capsule summaries (based on the 35-mm film format) are offered.

Fish-Eye   (less than 20 mm focal length). These ultra-wide-angle lenses are perfect for creating wonderfully artistic shots showing both the eclipsed Sun or Moon and earthly scenes, such as a distant horizon, the ocean, a cityscape, or a building or monument.

Although the Sun itself will appear very small, exciting images of a total solar eclipse's onset are possible by aiming the camera toward the zenith. By taking a series of exposures right around second and third contacts, the change in sky illumination and color will be dramatically recorded (figure 5.2) as the Moon's shadow approaches or recedes from your observing site.

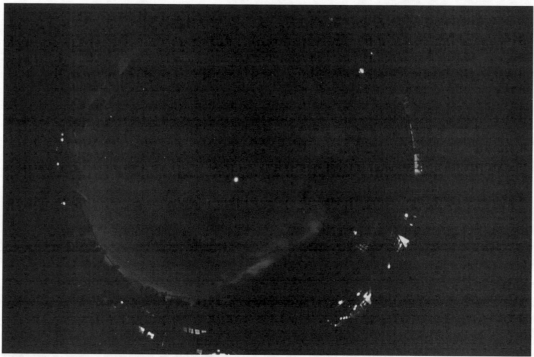

Figure 5.2 Wide-angle and fish-eye lenses offer some unique photographic possibilities, as evidenced by this photograph of the 1991 July 11 total solar eclipse. Photo by Spencer Rackley IV. (7.4-mm lens at f/4, 12-second exposure on Ektachrome 64 film).

Wide Angle (20 mm to 40 mm). Like fish-eye lenses, wide-angle lenses offer some wonderful possibilities for superimposing the eclipsed Sun or Moon in front of or over some aspect of the earthly environment. (Bear in mind that if you are shooting a partial solar eclipse, you will have to use a Sun filter to reduce the Sun's light to a safe, manageable level.)

Many photographers use wide-angle lenses to capture the eclipsed Moon among the surrounding stars and planets. This is perhaps the easiest eclipse shot of all to take successfully. Point the tripod-mounted camera toward the Moon, compose the shot to include other sky objects of interest, and shoot away. Be sure to use an exposure that is long enough to capture the fainter sky objects, as they will probably be fainter than the Moon itself. With ISO 400 film, a 30-to-45-second exposure at f/2.8 should be more than adequate. (See the discussion of film later in this chapter.)

Wide-angle lenses are also ideal for taking a sequence shot that records the entire event on one frame of film. By taking a photograph once every two to ten minutes, the Sun or Moon will appear to form a chain of images across the frame. Then, if you want, photograph a horizon along the bot-

Figure 5.3 The 1994 May 10 annular solar eclipse over Stellafane, the birthplace of amateur telescope making, in Springfield, Vermont. Photo by the author. (All solar images were made with a neutral-density 5 filter placed over a 50-mm lens. Early and late exposures were made at $\frac{1}{125}$th-second exposure and f/8, while the exposures at and around annularity were increased to $\frac{1}{60}$th-second at f/5.6; Kodachrome 64 slide film. After the eclipse, a normal, unfiltered exposure captured the foreground.)

tom of the frame to add interest. Figure 5.3 shows the 1994 May 10 annular eclipse over Stellafane, the birthplace of amateur telescope making in Springfield, Vermont.

To take a sequence shot on a single frame, it helps to use a camera designed to take multiple exposures. While comparatively few modern 35-mm cameras have this feature, non-motor-driven cameras can be fooled by pressing the film-spool release button (frequently found on the bottom of the camera's body) and holding the film-rewind knob while cocking the film-advance lever.

Plan the photograph before you make the first exposure. How high will the image be at maximum eclipse? Ideally, you will want that moment to be centered in the frame, so note that time and location, then count backward to when the first exposure should be made. If possible, run a *no-film*

Figure 5.4 A multiple-exposure sequence shot of the 1991 July 11 total solar eclipse, as photographed by Frans Pyck. (85-mm lens at f/5.6, ¹⁄₁₂₅th-second exposure on ISO 25 film.)

dress rehearsal before the eclipse. If you are photographing a solar eclipse (figure 5.4), do this a day or two beforehand; for a lunar eclipse, your dress rehearsal should take place a night or two after the previous Full Moon (in either case, you want the eclipsed object to be at approximately the same location in the sky as during the eclipse itself). Place the Sun or Moon, as appropriate, toward the eastern edge of the frame and estimate the path that it will travel across the sky. Then set the camera, sit back, and follow its track across the sky, keeping an eye on the time. If it doesn't work out, try again the next day. Keep at it until you are confident that you will get the desired result at eclipse time. Once determined, mark your tripod's settings and, if possible, its exact location on the ground. This way, come eclipse time, the camera can be repositioned exactly. Remember, above all, the camera must not be moved at any time during the eclipse, or the sequence will be broken.

Lunar eclipses also lend themselves to "Moon trails," such as that shown in figure 5.5. In this case, the tripod-mounted camera is aimed so that the Moon is positioned along one side of the frame. Position the Moon

Figure 5.5   A Moon trail, showing the evolution of the 1993 November 28 total lunar eclipse. By placing the camera on a stationary tripod, closing the lens to a small f/stop, and opening the shutter, the film records the changing appearance of the eclipsed Moon as the Earth's rotation carries it across the frame. (28-mm lens at f/16, Ektachrome 64 slide film, time exposure.)

so that, as it moves in the sky, it will creep across the photograph. If the Moon is rising, place it along the bottom of the frame; if setting, then put the Moon toward the top. Then close down the camera lens to its smallest f-stop and set the shutter to "bulb." With a locking cable release in place, start the exposure by depressing and locking the release's plunger. By keeping the shutter open throughout the eclipse, the Moon will appear to narrow, then broaden as it travels through Earth's shadow.

Normal   (45 mm to 55 mm). The lenses that come standard with most cameras are also useful for taking sequence shots as described above, but are somewhat more limited by their narrower fields of view. As a result, planning the shot becomes even more critical in order to prevent the eclipsed Sun or Moon from leaving the frame prematurely. Akira Fujii's stunning sequence shots of a total solar eclipse (front cover) and annular solar eclipse (back cover) serve as testimony to what can be done by the creative photographer.

Normal-focal-length lenses are also ideal for photographing the image of a partial solar eclipse projected onto some sort of screen. If you are viewing with a telescope or binoculars, be sure to set up a shade to help dampen out any stray light from washing out the Sun's image, as Massachusetts amateur astronomer Ed Faits shows in figure 5.6. And don't forget to look under nearby trees, whose intertwining branches and leaves act as

Figure 5.6 You don't need elaborate telescopes and expensive cameras to view and photograph an eclipse. For the 1994 May 10 annular eclipse, Ed Faits set up this small reflector to project the image of the Sun into a cardboard box. The clock at the lower right is useful for recording the exact instant each picture is taken.

nature's pinhole cameras to project tiny crescent Suns onto the ground, offering yet another photo opportunity, as seen previously in figure 3.2.

Moderate Telephoto   (85 mm to 400 mm). Perhaps the single most popular way of photographing eclipses is with a telephoto lens somewhere within this broad focal-length range. Moderate telephotos are ideal for making single exposures that record the progress of an eclipse. Their comparatively wide fields (in contrast to the longer focal-length telephotos described below) are well suited for capturing the full extent of the Sun's outer corona during the brief moments of solar totality, or recording any planets or bright stars that happen to be near the Moon during lunar totality. Simply set the camera on a tripod, aim, and fire.

Another fascinating type of lunar eclipse sequence photograph best taken with a moderate telephoto is something I call a "shadow sequence," as you can again see from Akira Fujii's beautiful lunar eclipse shadow sequence adorning the front cover. As before, these multiple-exposure photographs are taken on a single frame of film, but unlike the previous sequence shots, the camera is not held steady for a shadow sequence; it is moving. The camera is not tracking the Moon, however; it is following the *stars*. The net result is a picture that shows the Moon at several discrete stages in its silent journey through the Earth's round shadow.

A lunar eclipse shadow sequence is best taken with the camera and lens mounted piggyback alongside a telescope that is being piloted across the sky by a motorized clock drive. To ensure that the camera is tracking the sky properly, center a star in the telescope's field of view (preferably using an eyepiece with crosshairs). Don't be concerned if the star shifts in, or even leaves, the field of view; just make certain that it is always back in the center whenever an exposure is made.[1]

Symmetry is the key to a shadow-sequence photograph, so plan the number and times of all exposures accordingly. For a total eclipse, five individual exposures would seem appropriate, while three will suffice for a partial eclipse. You will want to take a photograph around mid-eclipse, so work the timing of the others out evenly from this point. For example, for a total lunar eclipse, take two exposures before mid-eclipse—say, 45 minutes and 90 minutes beforehand. Continue the sequence by making an exposure exactly at mid-eclipse, another at 45 minutes after mid-eclipse, and finally a fifth exposure 90 minutes after mid-eclipse.

Long Telephoto   (400 mm to 1000 mm) lenses are ideal for capturing the Sun's middle and outer corona, Baily's Beads, and the diamond-ring effect. This focal-length range is also well suited for partial and annular solar eclipses, as well as partial and total lunar eclipses. Long telephotos are likely the shortest focal lengths capable of recording surface details, such as sunspots and lunar craters.

Telescopes   (1000 mm and beyond). The longer focal lengths and greater apertures of telescopes allow the photographer to zero in on the full disk of the Sun or Moon during the various stages of an eclipse. Though a telescope's higher effective magnification (and resulting smaller field of view) may be inappropriate for capturing the gossamer veils of the outer solar corona, it is ideal for photographing the inner corona, chromosphere, graceful solar prominences (figure 5.7), and the ever-changing coloration of a lunar eclipse.

For photography, it is critical that the telescope be securely supported on a sturdy mounting, preferably one with a clock drive. Properly aligned equatorial mounts go a long way in easing the pressure of keeping a telescope aimed at the Sun or Moon. How to polar-align an equatorial mount,

Figure 5.7 The 1991 July 11 total solar eclipse displayed some magnificent prominences, as seen in this photograph by Frans Pyck. (100-mm f/5 refractor and a 2× tele-extender, 1/60th-second exposure on ISO 100 film.)

especially if you have traveled to an unfamiliar latitude, can be challenging, so be sure to review the procedure offered in Appendix F.

Lens Quality.   Regardless of the focal length of your telescope or telephoto lens, *never skimp on lens quality.* Lesser brands, though they may perform well under normal lighting conditions, are usually plagued by internal reflections and other optical flaws under the unusual circumstances of an eclipse. Fortunately, most lenses made by established companies (such as Nikon, Canon, Pentax, Minolta, and Olympus, as well as after-market brands such as Tamron and Sigma) are fully multicoated with a thin, antireflection coating of magnesium fluoride that dampens ghost images and flaring.

Don't go by name alone; always check out a new lens before an eclipse. Here's a test that never fails to show a lens's true colors, so to speak. Place the Full Moon slightly off center in the camera's field of view. Then, referring to the suggested exposures found in figure 5.12, take a series of three photographs as follows. With the lens aperture set fully open, take the first photo using the proper exposure read from the table. The second exposure should be increased by one stop (that is, the exposure is lengthened by one setting, such as from $\frac{1}{125}$th to $\frac{1}{60}$th of a second). Finally, overexpose the third photograph by two stops. If lens flare or "ghosting" is going to show up at all, you'll see it in these severe test photographs; otherwise, consider the lens to be well corrected.

Camera-Telescope Combinations

There are five basic camera-telescope combinations available to the astrophotographer: afocal, prime focus, positive projection, negative projection, and telecompression. Each is useful for photographing certain aspects of an eclipse.

Afocal.   The afocal system (figure 5.8a) uses both a camera's lens and a telescope's eyepiece. This allows just about any camera to be used for through-the-telescope photography, whether or not its lens is removable.

Before determining the proper exposure for your purposes, you'll need to figure out the effective focal ratio (EFR) of the camera-telescope combination. For the afocal method, the EFR can be calculated using the following formula:

$$\text{EFR} = (\text{camera lens focal length} \times \text{telescope magnification})$$
$$\div \text{ aperture of telescope}$$

Choose the eyepiece carefully, as it will determine whether all or only part of the Sun's or Moon's disk will fill the camera's field of view. For example, consider taking a photograph through a 20-cm (8-inch) f/10 telescope using a 25-mm eyepiece and a 50-mm f/1.8 camera lens. Dividing the eyepiece's focal length into that of the telescope determines that the system's magnification is 80 power. Plugging these values into the formula above yields:

$$\text{EFR} = (50 \text{ mm} \times 80\text{-power}) \div 200 \text{ mm} = f/20$$

With the f-ratio known, an estimation for proper exposure may now be selected from either figure 5.11 or 5.12, found later in this chapter.

The biggest problem in trying to photograph an eclipse afocally, especially a solar event, is preventing stray light from coming between the telescope and camera. Draping a black cloth over the camera and telescope, much as photographers did in the nineteenth century, will go a long way to dampen flare and light contamination.

Prime Focus   (figure 5.8b) is preferred by most photographers who wish to capture the entire solar or lunar disk in a single shot. This method couples the camera body directly to the eyepiece-less telescope, in effect turning the telescope into a large telephoto lens.

Determining the effective focal ratio requires no calculation; it is simply the telescope's own focal ratio. For example, the prime-focus EFR of a 20-cm f/10 telescope is f/10, while the EFR of a 10-cm (4-inch) f/10 telescope is f/10, and so on.

To take prime-focus photos through a telescope, the camera is attached directly to the telescope using a T-ring designed for your specific camera and a camera-to-telescope adapter (both are available from Orion Telescope Center; see Appendix A). Focusing is done directly through the camera's viewfinder by turning the telescope's focusing knob in and out until the image is sharp.

Positive Projection   (figure 5.8c), sometimes called "eyepiece projection," uses an eyepiece to enlarge the image of whatever is being photographed, in effect stretching a telescope's focal length. This makes it the method of choice if you are interested in capturing close-up structural detail in the chromosphere and prominences, or perhaps the Earth's shadow passing over a specific lunar crater.

As with prime focus, a camera-to-telescope adapter is used, but this time an eyepiece held in an extension tube is placed between it and the T-ring. The eyepiece projects the image from the telescope into the camera. Just how much the image will be magnified depends on the telescope's focal length, the focal length of the eyepiece, and the eyepiece-to-film projec-

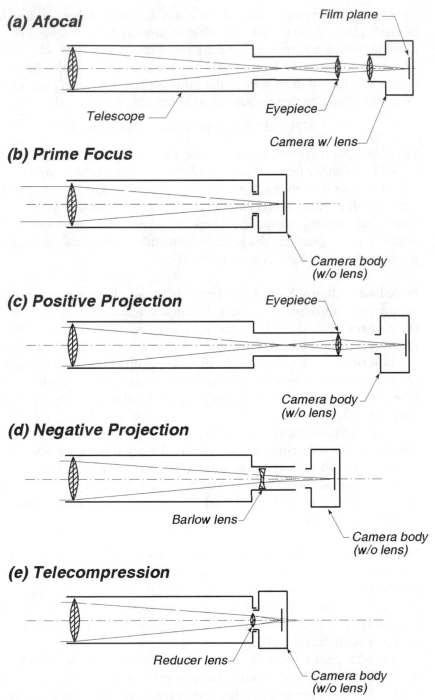

Figure 5.8 Five popular camera/telescope configurations.

tion distance. All other things being equal, the greater the projection distance, the higher the effective magnification.

As with the afocal and prime-focus systems, the effective focal ratio of the camera-telescope pairing must be known before proper exposure can be estimated. The following equation may be used for this calculation:

$$\text{EFR} = f_t \times [(L_e - f_e) \div f_e]$$

where:

f_t = telescope focal ratio
L_e = projection distance from eyepiece
f_e = eyepiece focal length

To illustrate this, let's go back to our 20-cm-aperture f/10 telescope. Suppose a 16-mm eyepiece is selected, with an eyepiece-to-film projection distance of 75 mm (3 inches). Plugging these values into the above equation produces:

$$\text{EFR} = \text{f/10} \times [(75 - 16) \div 16] = \text{f/10} \times 3.7 = \text{f/37}$$

This combination produces an effective focal ratio of f/37 and an effective focal length of 7,400 mm (2000 mm $\times$ 3.7). Exposure estimates may now be read from the forthcoming tables.

Negative Projection   (figure 5.8d) is similar to positive projection, except that a negative lens is placed between the telescope and the camera body. Negative projection does not extend a telescope's focal length as much as positive projection potentially can, making this method more ideal for photographing broad portions of the Sun's or Moon's limb (such as when photographing solar prominences).

The most commonly available negative lenses are the Barlow lens and the photographic teleconverter. Many photographers prefer the latter. To use a teleconverter on a telescope, simply connect it to the camera body and use a camera-to-telescope adapter like that used for prime-focus work. A Barlow lens may be used as a projection lens by inserting it into the telescope's eyepiece holder and attaching the camera to it using a camera-to-telescope adapter and a T-ring. The resulting increase in magnification will be equal to the power of the teleconverter or Barlow, usually either 2× or 3×. Therefore, using either a 2× teleconverter or a 2× Barlow with, for example, an f/10 telescope will double both the instrument's focal length and effective focal ratio (in this case, to f/20).

Telecompression   (figure 5.8e) may be thought of as reverse projection. Rather than enlarging the image, a telecompressor or focal reducer placed between the telescope and the camera body will lower the effective focal ratio. The net result is a negative magnification effect and a wider field of view. This is especially useful when trying to photograph the full disk of either the Moon or Sun through a long-focal-length telescope.

Most telecompressors/focal reducers state their "deflation factors" right on them. For instance, many cut the effective focal ratio of an f/10 telescope by 37 percent to f/6.3 (and, therefore, the effective focal length from 2000 mm to 1250 mm, in the case of a 20-cm f/10 instrument), while others have a reduction factor of 50 percent (turning an f/10 to an f/5, for instance). Just be sure that the telecompressor you choose will correct for any distortion that will occur around the edges of the film frame. And don't forget to adjust your exposure estimate accordingly.

Film

Choosing the right film is just as important as choosing the right lens. Today's photographers are both blessed and cursed with a great variety of film from which to select. An eclipse is not, however, the time to test out a new film; those tests must take place long beforehand.

Should you take slides or prints? Both offer pros and cons. Slide films eliminate one critical step in the subject-to-finished-photo process—printing. Anyone who has ever taken color prints of just about any subject, astronomical or earthly, knows that poor quality control at the finishing laboratory can turn a first-rate negative into a muddy, ill-focused, improperly exposed print. Slides eliminate that error factor, and so are preferred by many photographers.

Slides are not without their downside, however. Frequently, photographers will speak of a film's *latitude*. In this context, latitude refers to a film's forgiveness factor, or just how far off an exposure can be and still get a reasonable end result. Here, color print film wins handily over slide film. Even color negatives over- or underexposed by as many as two, three, or more f-stops can still produce acceptable color prints, given a skilled darkroom technician.

The implications of this are great to the eclipse photographer. Not only will the wide latitude of print film let you get away with improper exposures during the partial eclipse phases, it will also let you record a wide range of details during totality. Depending on how it is printed, a single color negative will produce good-quality prints of the Sun's chromosphere, prominences, and inner and middle corona, or the eclipsed and uneclipsed portions of a lunar eclipse. To produce a similar array of photos on slide film, a photographer would have to take two, three, or more different exposures, each set for the phenomenon of interest.

Keep in mind also that to make a print from a slide usually requires that an *internegative* be made from the slide first, likely resulting in the loss of fine details. It is also possible to produce a print directly from a slide, but this can result in artificially increased contrast. Color negatives can be directly turned into slides by copying them onto another color negative film, though with some increase in contrast as well.

What about film speed, or ISO (formerly ASA)? As a general rule, the slower a film's speed (the lower the ISO number), the finer its granular structure, or grain, and the greater its sharpness. These are important considerations, especially if you will be projecting the slides onto a large screen or making enlargements from the negatives.

To add intrigue to the question of which film is best for eclipse photography, film manufacturers introduce new emulsions faster than most people can shoot them. For solar-eclipse photography, some of the better films at present include Fujichrome Velvia (ISO 50), Kodachrome 64 and 64 Pro, Ektachrome Lumiere 100, and Fujichrome Provia and Sensia 100 Pro slide films; and Kodak Royal Gold 25, Fujicolor Reala (ISO 100), Kodak Royal Gold 100, and Fujicolor NPS 160 color-negative films. If you are photographing the eclipse from a ship, airplane, or other moving vehicle, selecting faster (higher ISO value) film will allow shorter exposures to minimize any blurring. Recommendations here include Kodak Ektachrome 200 Pro and Fujichrome Sensia 200 RM slide films, as well as Fujicolor Super G 200 and Super G Plus 400, and Kodak's Royal Gold and Ultra 400 color-negative films.

Selection of film for photographing a lunar eclipse is a little trickier. For the partial phases, the advice given above about using a slow, fine-grain film applies here as well. During totality, however, the Moon's disk may drop in brightness from magnitude -12 to magnitude $+4$, a reduction factor of 2.5 million times! This tremendous range makes slower films impractical. To offset this large light loss, experienced lunar-eclipse photographers, even those with telescopes held on very sturdy, clock-driven equatorial mountings, prefer faster films, such as the ISO 200 and ISO 400 emulsions noted above. Even then, exposures of several minutes might be needed to record the eclipsed Moon's image successfully.

This short list of films is by no means exhaustive, and indeed may be out of date by the time you read this! For late-breaking, up-to-the-minute film evaluations, consult recent issues of photography magazines such as *Popular Photography* and *Outdoor Photography*. Though they do not specifically evaluate a film's usefulness for photographing eclipses, these magazines will discuss important criteria such as grain structure, exposure latitude, and color sensitivity.

Tripods

Tripods are critical to the successful eclipse photographer. Many of the less expensive tripods sold in department stores and other mass-market outlets

Figure 5.9 No, this "Moon trail" wasn't taken during an earthquake, but it is a good indication of what can go wrong when you use a shaky tripod!

are just not sturdy enough to support a camera steadily. If the tripod is shaky, then the photographs will be hopelessly blurred. Want proof? Take a look at figure 5.9, which shows a jittery "Moon trail!"

Make sure your tripod fits the occasion. The legs should be extendable so that the camera may be raised to a comfortable height, and strong enough to remain steady. Better models use braces between the tripod's legs and the center elevator post. These same units will also likely have convertible footpads that feature both a rubber pad, for use on a solid surface, and spike, for softer surfaces like grass or dirt.

Review the discussion on telescope mounts in chapter 2 for additional thoughts.

Cable Releases

Every time you depress the shutter release button, you chance bumping the camera and blurring the photograph. That's why, especially for long focal-length shots or long exposures, it is strongly recommended to use a cable release. This is a wire plunger that typically screws into the shutter button. Instead of depressing the shutter button directly, the photographer pushes the plunger to trigger the exposure. But heed this warning: cable releases always seem to break just when you can least afford it, so always bring along a spare.

Motor Drives

Another secret of the successful eclipse photographer is to do as little thinking as possible during an eclipse, especially during the brief plunge into solar-eclipse totality or annularity. Those moments are so precious that you want to automate as much of your equipment as possible.

One of the easiest ways to help ease your anxiety during an eclipse is to add a motor drive to your camera. Motor drives are battery-powered attachments that automatically advance the film to the next frame immediately after a photograph is taken. Many of today's cameras come with motor drives built right in, while others may be mated to the bottom of the camera.

While motor drives make life a little simpler for the eclipse photographer, keep in mind that they also vibrate during operation. Therefore, resist the urge to fire away rapidly. Wait about three to five seconds between exposures, so that any vibrations will dampen.

Right about now, solar-eclipse photographers may be thinking, "But five seconds is an eternity during totality." True, it seems a long time when you have to count it off during such a nerve-racking event, but which would you rather have—thirty motion-blurred images or half a dozen razor-sharp photographs?

Focusing Screens

Should you own one of the relatively few cameras designed with removable focusing screens, it is best to replace the standard screen with one of clear glass. You will find it much easier to focus (and just see) your subject.

Video Cameras

What about videography of eclipses? With their prices dropping, and features and sensitivity rising, family-style video camcorders offer excellent ways of capturing the magic and majesty of an eclipse. But which format is best: full-size VHS, VHS-C, or 8 mm? Full-size VHS offers several distinct advantages, not the least of which is that the tapes are readily available anywhere from pharmacies to gift shops, while other formats can prove more difficult to find. In addition, full-size VHS tapes may be played directly in just about any VCR without copying or using a separate adapter. On the plus side of VHS-C and 8 mm camcorders are their tiny sizes and lighter weights, making them easier to use and simpler to mount along the side of a telescope or other support. In addition, many videographers feel that 8 mm cameras have better resolution than the others, an issue that is often debated with a fervor equal to the greatest political and religious arguments. Regardless of format, there is no denying that video resolution is not as high as that of film.

While we continue to hear wonderful stories about the coming of so-called high-definition television (HDTV), the fact is that its arrival in the United States and Canada is still years away. Thankfully, though most of us are still slaves to comparatively archaic television technology, higher-resolution camcorders are here today. Truth be told, ordinary VHS and 8-mm camcorders can record eclipses in adequate detail, but for outstanding results, you just can't beat Super-VHS (S-VHS) and Hi-8 models. The continued growth in popularity of these next-generation camcorders offers great promise for recording even greater detail, but keep in mind that the image quality will only be as good as the VCR you are using to play it back.

In addition to format, here are some other points worth consideration. First, let's examine image size. Just about all camcorders sold today feature zoom lenses with focal lengths between 6:1 (6×) and 14:1 (14×). While these are adequate for photographing a total, annular, or partial solar eclipse, longer focal-length lenses are preferred. Some high-end camcorders, such as the Canon L2, even feature interchangeable lenses, greatly expanding their creative capabilities.

Though less expensive camcorders come with nondetachable lenses, their effective magnification range can be extended with a supplemental tele-extender. These screw onto the front of the camera's lens and effectively double or triple its focal length. Just one word of caution when using these: some cheaper tele-extenders are made with poor-quality optics that can produce flare. Only consider tele-extenders with multicoated lenses.

With the popularity of solar eclipse cruises, videographers might consider a camcorder with a built-in image stabilizer. An image stabilizer is designed to smooth out the jiggles and wobbles often associated with hand-held shots. Depending on the sea's rocking motion during the eclipse, these may also help to smooth out a shipboard video. Keep in mind, however, that image stabilizers are of limited effectiveness (some brands more so than others), and that a sturdy, land-bound tripod is still best for steady shots.

Finally, to capture the greatest amount of detail, use the highest-grade videotape you can find. Better-quality name-brand tapes, such as TDK, Kodak, and Maxell, will always yield noticeably better definition than so-called bargain "high standard" tapes. After all, you're not taping your favorite television show! You're recording an eclipse.

Solar Eclipse Specialty Items

Here are a final few photographic accessories that you might consider using to capture different aspects of a total solar eclipse. Each is recommended only for veteran photographers who already have several notches on their total-eclipse belts.

Diffraction Gratings. Of the great variety of filters and other lens attachments available in today's photographic marketplace, one that offers some interesting possibilities is the diffraction grating. Diffraction gratings are thin pieces of clear acetate that have been etched with a series of tightly spaced grooves—some 750 per millimeter! These grooves, or lines, bend and spread (or diffract) light rays into their component colors, or spectrum.

If you hold a diffraction grating up to an incandescent light, you will see a continuous spectrum, ranging from red to violet. If you then hold the grating up to a fluorescent light, you will see not a continuous spectrum at all, but an emission-line spectrum showing the bright lines of the element mercury instead.

A similar effect can be observed during a total (and *only* a total) solar eclipse. By viewing the chromosphere and corona through a diffraction grating, we can discover the characteristic spectral emissions of each, in what is called the *flash spectrum.*

To photograph the flash spectrum, you need only place a piece of diffraction grating in front of your camera lens. Handle the grating by the edges only, because oils from the skin can damage the film's properties. Better still, mount the grating in a cardboard slide mount or similar frame, and then tape the mount to the front of the lens. Once secured, aim the camera toward the Sun, but just before second contact, as diffraction gratings offer no protection against (and, in fact, can be damaged by) the intense rays from the partially eclipsed Sun. As totality begins, position the Sun in one corner of the field and follow the exposure recommendations offered in figure 5.11. When processed, the photographs will show both the eclipsed Sun and its spectrum vividly displayed across the frame.

Exciting results are possible by placing a diffraction grating in front of a video camera and letting the tape roll in real time (i.e., not time-compressed) as totality begins. If the Sun is positioned in the frame correctly, the tape will record the changing appearance of the continuous spectra of Baily's Beads and the diamond ring, as well as the line spectra of the chromosphere and corona.

Diffraction gratings are available commercially from a number of sources, including Learning Technologies and Edmund Scientific Company (see Appendix A). So-called holographic gratings will produce brighter images, and are therefore recommended for photographing the flash spectrum.

Radial-Gradient Filters.   These filters vary in density as the distance from the center increases. The idea is simple enough: the denser central portion of the filter suppresses the brighter chromosphere and inner corona, while the thinner portions near the edge pass light from the outer corona without obstruction.

Radial-gradient filters are not sold commercially. Instead, they must be custom-made to a specific camera-lens combination. The process is an involved one, and as such, it is beyond the scope of this book. It should be noted that many photographers are now favoring darkroom techniques for producing similar results achievable with radial-gradient filters. These are discussed later in this chapter.

SETTING THE MOOD

You are all set to photograph an eclipse. It doesn't matter if it is a solar or lunar event, or a penumbral, partial, annular, or total eclipse; all require forethought. The time to plan your photographs is not during the eclipse itself; the time to plan is NOW, weeks, even months beforehand.

Focusing

Regardless of the system, focusing is absolutely the key to success and is, ironically, probably the least considered item by first-time astrophotographers. After all, you just turn the lens focusing ring to infinity (∞) and forget it. Right? Or, better still, let the autofocus feature do the focusing for you. What could be easier?

Ahh, if only it were that easy. Determining your lens's precise focus point, especially with a telephoto lens, is critical to a photograph's quality. The focus markings may be off enough to turn an otherwise properly composed and exposed photo into a blurry mess. Worse yet, if you are shooting through a telescope, there are no focus markings; these have to be determined from scratch. To add to the confusion, a glass solar filter may change the focus point. The bottom line to all this is never, ever to trust a lens's focus markings; determine it for yourself. (Focus the image with your eyeglasses on, should you wear them.)

Here's an easy way to get a sharp focus the first time, every time, on the Sun or Moon. Long before eclipse day, make a front-end mask for your telescope from a circular piece of cardboard that is the same diameter as the telescope's (or telephoto lens's) aperture. In the mask, cut two smaller circles directly opposite each other. (Five-centimeter diameters work well for 20-cm and larger telescopes.) If you are working with a reflecting or catadioptric telescope, or a mirror telephoto, make certain that the holes are not blocked by the instrument's secondary mirror.

Attach the mask in front of the telescope (and solar filter, if viewing the Sun), mount the camera, aim at the target, and look through the camera viewfinder. If the telescope is out of focus, you will see two overlapping images. Turn the focusing knob in or out until the two images slowly blend into one, as shown in figure 5.10. When a single image is seen, the telescope is properly focused. Once set, mark where the focus point is on the instrument's drawtube or barrel, if possible, so that the focus setting can be easily repeated.

A commercial version of this device, called the Kwik Focus, is marketed by P&S Sky Products. Other, more elaborate (and therefore costlier) focusing devices for telescopes include the SureSharp and PointSource, both by Spectra Astro Systems, and Celestron's Multi-Function Focal Tester (see Appendix A).

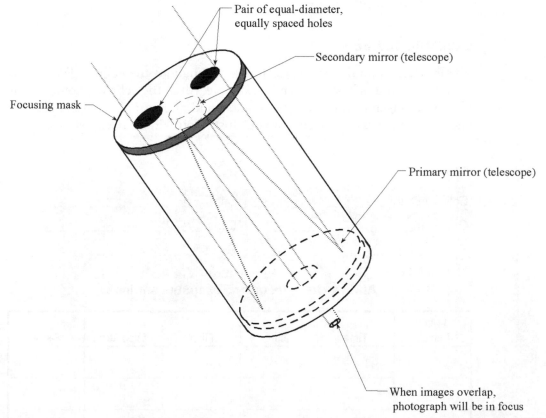

Figure 5.10 A focusing mask placed over the front of a telescope makes it easy to check that an image is sharp and clear before taking an exposure.

Keeping Count

There is nothing worse than running out of film just before maximum eclipse! A word to the wise: plan out your exposures beforehand. Complete a photographic itinerary such as that in table 5.1, and then stick to it regardless of how much adrenaline is running amuck in your body on eclipse day. Avoid the urge to fire at will. The successful photographer always knows exactly when a film change is coming up.

Continue the table for the full length of your particular roll of film. Note that most rolls allow for a couple of extra shots at the end, so add a few extra lines just in case.

After you advance the film, always pause for vibrations to settle out. Allow three seconds for short exposures (say, $1/250$ and faster) and five seconds for longer exposures. It's always better to get a few sharp pictures of the event than a multitude of blurry images. (If you want to take more photographs than one roll will permit during a total or annular solar eclipse, bring along another camera. Don't make the mistake of trying to change rolls of film during totality!)

Exposing Yourself

Finally, determine the exposures to use throughout the eclipse. I've always been told that it is best to learn by example. All of the eclipse photos in this book were taken by amateur astronomers and photographers—just like you and me. If they can do it, so can you! Just about all of the exposures, f-stop

Table 5.1

Eclipse Date: _____   Site: _____

Camera: _____   Lens: _____

Film: _____   Roll # _____ of _____

Remember!
Always check focus and exposure settings.

Frame Number	Time	Film	Filter	Exposure	F-stop
1					
2					
3					

settings, and films used for these photos are specified in the captions. Find one or more that was taken with equipment that is similar to your own, and study the results. If it's the kind of picture that you would like to emulate, make a note, and then practice for your time in the shadow.

If you wish, instead, to venture into new waters and try something different, so much the better! Figure 5.11 offers suggested exposures and f-stop settings for just about every stage of a solar eclipse, while figure 5.12 lists exposures for lunar eclipses. Remember, these are only meant for a *general* guide, simply a good place to start. The altitude of the Sun or Moon and sky transparency are just two of the factors that can affect exposure. Don't be afraid to bracket the exposures up and down by one or two f-stops from these recommendations, and, by all means, always take a test roll well before the eclipse.

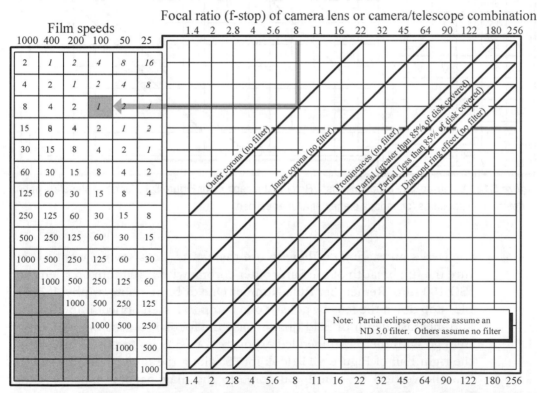

Figure 5.11 Solar-eclipse exposure chart. Along the top of the table's left-hand side are ISO speeds of films, such as ISO 50, ISO 100, and so on. Below these are numbers representing shutter speeds. Italicized numbers are exposures in whole seconds, while the others are in fractions of a second (e.g., *125* translates to ¹⁄₁₂₅ of a second, while *2* means 2 seconds). Along the top of the right-hand side of the chart are various lens f-stop settings, such as f/2.8, f/4, and f/11. Beneath these is a series of diagonal lines labeled with various stages of a solar eclipse.

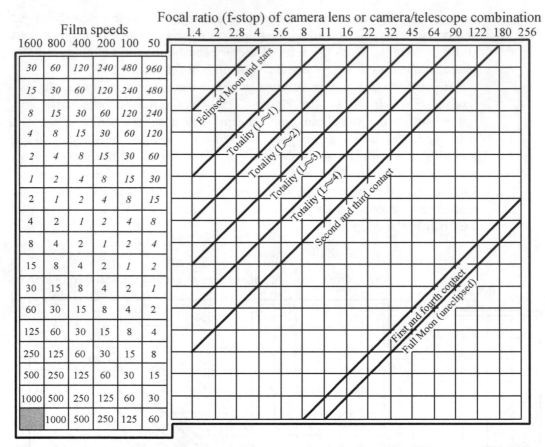

Figure 5.12   Lunar-eclipse exposure chart. Refer to the caption for Figure 5.11 for an explanation of how to determine exposures for various stages of a lunar eclipse.

Here's an example of how to use the charts. Let's say we want to capture the full breadth of the Sun's outer corona with a 400-mm f/5.6 lens and ISO 100 film. Remember Rule Number One of basic photography: If possible, stop down the lens by at least one f-stop to lessen optical imperfections and increase depth of field (just in case the focus is a little off). So, rather than go with the recommended exposure for f/5.6, go to the f/8 column on the right half of Figure 5.11 (look for "8" along the top or bottom) and follow the vertical line until you come to the diagonal line labeled "Outer Corona." Turn left and trace the horizontal line from that intersection point into the chart's left-hand portion. Each column there is labeled with a film speed (ISO value). Find the column labeled "100" and trace it down to where it meets the horizontal line you were following from the right. That's your exposure; in this case, 1 second.

The diamond ring is perhaps the single most spectacular part of a total solar eclipse. Capturing it requires speed, dexterity, a cool head, and some good, old-fashioned luck! First you must be prepared to remove the protective solar filter just before the very instant of totality—too soon, and the Sun's strong rays flood the camera's sensitive interior (to say nothing of your eyes), but too late, and you miss the show! The confusion continues with exposure selection. Though the chart suggests $1/500$ for this lens/film combination, consider this as a minimum. In other words, good results may be achieved with longer exposures as well. Photograph the second-contact diamond ring at the given exposure, then increase that exposure by two f-stops (e.g., increase exposure to $1/125$ at f/8) to capture the diamond ring at third contact.

What if you're shooting with a 10-cm f/9.8 refractor, 20-cm f/10 Schmidt-Cassegrain telescope, or some other instrument with a focal ratio not listed on the table? No problem. Just choose the standard f-ratio that's closest. In the case of these two instruments, use the recommendations for f/11.

Fill in each value on the exposure table a day or two before the eclipse, remembering to bracket exposures by one or more f-stops just to make sure that you get at least one set of properly exposed photographs.

Captain Video

As previously mentioned, camcorders offer some outstanding possibilities for photographing eclipses that were unavailable even twenty years ago. Back then, if you were an amateur wanting to capture the live action of an eclipse on film, you probably used an 8-mm or Super 8 movie camera. Since most of these cameras were soundless, you had to record people's reactions on a separate tape recorder, then play both back in sync to capture the magic. All this for three and a half minutes of film. Of course, more ambitious moviemakers could shoot with 16-mm film, but the expense of the equipment was often prohibitive.

Enter the marvel of the modern-day video camcorder. Nowadays just about anyone can take high-quality, live-action footage of just about any event worth remembering, from little Helen's first-grade play and Brian's graduation to, yes, even eclipses.

Some cameras let the user vary the exposure, while others are designed to be "idiot-proof" and only feature automatic exposure control. To record the progress of a lunar eclipse, the videographer has to outsmart the brain of the camcorder, which is often confused at seeing this bright circle in the middle of an otherwise black field. Even at higher shutter speeds, a distinctive vertical "beam" is often projected above and below the image because of CCD-chip overload.

To eliminate this problem, try placing a single piece of overexposed black-and-white film in front of the lens, effectively creating an ND-2.5 filter. Then, as the eclipse progresses, remove the filter and adjust the image using the camera's high-shutter settings. By the time totality strikes, you might well be engaging the camcorder's low-light settings.

Solar eclipses can be videotaped in much the same way. We have already seen how a solar filter with a neutral-density value of 5 works well for both still photography and visual use. Camcorders on the other hand, even those with high-speed shutters, often yield better-exposed images with a higher neutral-density value. In such cases I have had good luck using three layers of fully exposed and developed black-and-white negatives to create an ND-7.5 filter.

If you will be videotaping a total solar eclipse, remember that you must leave the filters in place until only a few seconds before totality. You might want to leave one filter in place even during totality to record the inner corona and chromosphere, then remove it to register the outer corona.

Pre-eclipse equipment testing is a must, but how? Fortunately, the Full Moon is approximately the same brightness as the Sun's corona, so use it to determine the combination of filter and shutter speed that gives the best results. If it is necessary to use one of the filters even during totality, be sure to mount it separately, behind the others. That way, it will be easier to remove the ND-5 filter while leaving the ND-2.5 filter in place.

There is nothing duller than watching a videotape of an eclipse in real time. I get bored just writing about it! I strongly recommend that, rather than just aiming the camcorder skyward and letting the tape roll, you plan exposures just like a still photographer. Take a five-second clip every two to five minutes, or better yet, if your camcorder has a built-in time-lapse feature (intervalometer), let it do the work for you. A one-second clip taken every thirty seconds will reduce the hour-long partial phases leading up to totality down to a manageable two minutes (though even that is long).

Time-lapse photography of an eclipse requires careful tracking of the Sun or Moon to prevent its disk from bouncing wildly across the television screen. The best approach is to mount the video camera on a clock-driven equatorial mount. Of course, the effectiveness of this method depends entirely on the precision with which the mounting has been aligned to the Celestial Pole (see Appendix F).

Stake Your Claim

One final thought: Where are you going to observe from? Of course, you know the location, but exactly where? On what physical piece of terra firma will you be setting up? If you're on a chartered expedition and this is your first time at the observing site, scout out the area. A smooth, flat, grassy area is best. If possible, avoid setting up on blacktop, concrete, or other paved surfaces, because all of these create uncontrollable heat currents that can interfere with the clarity of photographs.

Try to plan ahead for every contingency by reviewing chapter 2 (especially table 2.1). For instance, if the area might be sandy, the tripod's legs will sink into the ground. To prevent this from happening, place a plastic coffee-can lid, paint-can lid, or even an upside-down Frisbee under each footpad to distribute the weight over a larger area. As previously noted, always bring along a white sheet to set all your equipment on. You'll find the sheet makes it easier to find anything that may drop.

DARKROOM WIZARDRY

Taking the photograph is only half of the story. Now it's time to get it developed. Whether you should process your own film or have it done by a commercial laboratory is up to you. If you take your film to a commercial laboratory for processing, tell them *not* to cut the film. There have been many instances when cutting machines, unable to find the edges of the frames owing to the dark subject matter, have inadvertently sliced perfectly good negatives and slides in half! It is best to bring the uncut film home and determine for yourself where one frame ends and the next begins. Alternatively, take a normal photograph on the first frame to give the processing machine a starting point to reference.

Getting a decent print of an eclipse negative or slide from a mail-away mass-market commercial laboratory, on the other hand, might prove difficult. If you have shot color-negative film, better results frequently come from taking a roll of film to a local 60-minute processor where the film is developed and printed right in the store. Go in during a slow period, and tell the technician to print the negatives so that the background appears jet black.

Some people choose to bring their negatives and slides to a custom photo laboratory. There a skilled darkroom artisan can work with each print individually until the desired result is achieved. Custom laboratories are not inexpensive, however, usually charging several times what your corner pharmacy would for an enlargement.

A few years ago, Kodak introduced the Creation Station and Copyprint Station machines. These clever, do-it-yourself devices let photographers make their own instant enlargements in a store while they wait. Contrast and brightness controls allow some customizing of the results before the "print" button is hit.

Then there are those who have access to complete color darkrooms. These individuals stand the best chance of getting optimum results from their original negatives or slides, for they can work each print until it is absolutely perfect. Not an easy task by any means, but by following a few of the tricks discussed below, some outstanding results are possible.

As mentioned earlier in the discussion on film selection, the broader exposure-latitude range of color negative film can yield several prints from a single shot, with each showing a different facet of the photographed image. A single negative of a total solar eclipse, for instance, may produce striking shots of the outer corona, inner corona, and prominences just by varying the time that the photographic print paper is exposed to the enlarged image. Short enlargement times will pick up the outer regions of the corona, while printing the negative longer will "burn in" the details closest to the Sun's eclipsed limb.

Dodging

Although the human eye is able to view all "layers" of an eclipse simultaneously, a photograph usually displays only one. A darkroom technique called *dodging* can, however, greatly increase a print's range by lightening or darkening small sections. To do this, photographers use dodging tools to alternately expose and hide critical areas of a negative during the printing process.

Dodging tools come in two varieties: *occulting disks* and *masks*. An occulting disk looks something like a wand, with an approximately circular piece of cardboard taped to a straightened length of wire coat hanger. A mask, on the other hand, is simply an irregular circle cut into the center of a sheet of cardboard.

Which tool should be used when? If a small area of a negative is especially dense, use a mask to hide the rest of the print while the enlarger burns in that portion. A disk is used for the opposite effect, that is, to hide a small portion of a negative while the enlarger exposes the remainder. In either case, the tool must always be kept moving during the printing process to prevent any distinct marks or edges from forming on the print.

Unsharp Masking

Another, more elaborate darkroom technique used to bring out a wide range of subtle details in a negative is called *unsharp masking*. If only a small area of a negative requires some special work, then dodging is fine, but for whole-negative work, unsharp masking produces far better results. How does it work? Imagine that you have a negative with a dark blob in the middle surrounded by some faint, tenuous detail (remember, in a negative, dark corresponds to a bright image on a print). The uneven image usually results in a print that is either too dark to see the tenuous detail or so bright that the object in the center overwhelms the view.

That's where an unsharp mask comes in. If you have worked in a darkroom, you are likely familiar with the term *contact print,* used to describe a print of a negative made by sandwiching the negative against a piece of photographic paper and exposing it to light. The result is a print the same size as the original negative.

You make an unsharp mask in exactly the same manner, but instead of making a contact print on photographic paper, you use a piece of unexposed film instead. When that piece of film is developed, the image turns into a "negative negative," or a positive, of the original. By holding both the original negative and the "negative negative" (i.e., the positive) together, you'll see that the "negative negative" more or less cancels out the burned-in areas of the original while letting the thinner areas come through un-abated.

This same technique can be used to help balance the printing of a too-contrasty negative, provided the image on the "negative negative" (i.e., the mask) is blurred just enough to soften any fine detail. Most prefer to slip a thin piece of glass (no more than a couple of millimeters thick) between the negative and the film, then expose the sandwich briefly to light from an enlarger. Exact exposure times are hard to predict, so expect some trial and error. Though conventional film can be used to make an unsharp mask, this will require that you work in total darkness. A better approach would be to use a film with a photographic-paper type of emulsion that can be handled under a darkroom's safety light, such as Kodak's 5302 Fine Grain Positive Film. Once the unsharp mask is made, overlap it and the negative in your enlarger's negative holder, and you are set to print.

Duplication

What if a negative or slide is incorrectly exposed or too low in contrast? Or what if a shot is not composed correctly? Or what if the image, though as sharp as a tack, is just too small? To remedy such problems, many photographers duplicate the original onto a second roll of film, using a slide copier. Although slide copiers come in many different shapes and sizes, the most common (and least expensive) consist of long tubes that attach to your camera. At the opposite end is a clip designed to accept a standard two-inch-by-two-inch slide mount. Inside the tube, a set of close-up lenses brings the slide into focus. Illumination comes from light diffusing through a piece of ground glass mounted in front of the slide clip. Some slide copiers require a separate flash, while others let you use daylight to light the slide from behind during the exposure. The latter are usually easier to use.

Because of the close range and limited depth of field, precise focusing is critical. Even a slight shift caused by a warped slide mount will blur an otherwise sharp image. That's why many experts recommend mounting the originals in glass slide mounts to keep them flat.

Correct color evaluation is also crucial for proper duplication. Kodak's Publication R-25, *Kodak Color Print Viewing Filter Kit*, is especially helpful for determining the exact hue that is present, and which color-compensating filter(s) might be required to retain or correct it. This publication (cross-referenced as catalog number 1500735) may be ordered from Silver Pixel Press (see Appendix A).

Nowadays, to create a simple duplicate slide, most people use Kodak Ektachrome Duplicating Films. When properly exposed and color-balanced, these films produce duplicates that are extremely close to the originals in every respect. If you are interested in increasing contrast in a shot, then try a regular daylight transparency film, such as Kodachrome 25. Some photographers copy a slide several times, with each succeeding generation gaining a little contrast, though frequently at the sacrifice of color. If you are interested in creating a color negative from a slide, use a film designed for the task, such as Kodak's Vericolor Internegative Films. Finally, to create color slides from color negatives, try Kodak's Vericolor Slide Film.

There are so many intricacies to duplicating and copying photographs that an entire book could be written on the subject. Fortunately, one has been: *Copying and Duplicating in Black-and-White and Color*, co-authored by W. Arthur Young, Thomas Benson, and George Eaton. This publication (cross-referenced as catalog number 1527969) may also be ordered from Silver Pixel Press.

Computer Processing

One final approach to darkroom wizardry does not require a darkroom, at least not in the conventional sense. Too much or too little contrast? Looking to pull out some latent detail? No problem! Enter the computer. Digitized photographic images can be electronically viewed and manipulated on a computer. Some better-equipped computer "darkrooms" have apparatus that can digitally scan a print or slide directly into the computer, but even that is no longer a requirement. For a fee, Kodak will transfer your color negatives onto a Photo CD-ROM that may then be viewed with appropriate software.

Wondrous results are possible with computer processing. One of the pioneers in this field is Steve Albers of Boulder, Colorado. To re-create the full visual range of the July 1991 total solar eclipse (figure 5.13), he used a VAX 6420 computer to combine five individual photographs, each exposed correctly for a different radius in the solar corona.

Figure 5.13 Computer meets camera in this digitally combined and enhanced photograph of the 1991 July 11 total solar eclipse by Steve Albers. Albers combined five individual photographs, each exposed for a different radius in the solar corona, to create an accurate portrayal of the eclipse's visual appearance. The inner four photographs were taken by Dennis di Cicco of *Sky & Telescope* magazine using a 1500-mm lens and medium-format Kodak Ektar 25 film. The fifth photo, showing the outer corona, was taken by Gary Emerson of E. E. Barnard Observatory using a 500-mm f/8 telephoto lens and Fujicolor 100 35-mm film. These photographs were digitized at the National Center for Atmospheric Research's High Altitude Observatory by David Sime using a CCD scanner. Albers then processed the digital data using complex image-processing software that he designed specifically to combine the digitized images while also enhancing coronal detail.

Albers describes the algorithm he developed: "The processing included many steps that started with registering the images [i.e., ensuring that all five images were aligned with one another]. A digital mask was then constructed to isolate the optimally exposed region in each photo, creating, in effect, a set of five annular-shaped regions of the corona. Each image was then given a contrast stretch to correct for the film's response and exposure level. This stretch was varied slightly to assure a seamless fit after the images were combined. After some tinkering, I had a pleasing image that shows the full dynamic range of the corona." Some tinkering? A mild understatement, since the project consumed some 500 hours across an 18-month period. Albers's techniques were described more fully in Dennis di Cicco's article "Eclipse Photography Goes Digital," which appeared in the November 1994 issue of *Sky & Telescope* magazine.

That same article also featured the digital technique of Kazuo Shiota of Japan. Using the popular graphics program Adobe Photostyler, Shiota was able to combine six eclipse images (ranging in exposure from $\frac{1}{250}$ of a second to 4 seconds on Fujichrome Velvia film) into a full-range photograph. Like Albers, he digitally produced an unsharp mask to isolate only a portion of each shot, then combined them into a single image. Shiota's efforts were also very time-intensive, taking six months to get the image just as he wanted it.

Most experts agree that getting a good color print from a computerized image is difficult. Though technology is always changing, at present the finest prints are made with a thermal dye diffusion digital printer. With the best printers, it is often difficult to tell a computer-generated print from a true photographic enlargement.

These early attempts have concentrated on revealing hidden details in solar-eclipse images, but similar results showing both the eclipsed and uneclipsed portions of the Moon during a lunar eclipse are also possible. If you have a penchant for computer "hacking," this might be a great way to express your creativity and your fluency in "computerese."

All this is on the cutting edge of technology today, but will undoubtedly be considered old hat long before many of the eclipses detailed in this book are ever seen. Endless exciting possibilities in the world of electronic photography and imaging software are developing (no pun intended) even as this is being written. Charge-Coupled Devices (CCDs), once only in the realm of professionals, are now used routinely by amateur astronomers to photograph a wide range of celestial sights, from the planets to distant galaxies. Unfortunately, most CCD cameras in use as this book is published suffer from very small fields of view, owing to the small sizes of their CCD chips. While this may make them too restricted for most aspects of eclipse photography, advances in technology will soon make new, wider-field CCD cameras readily available.

But, regardless of how you choose to photograph an eclipse, either electronically or with conventional film, let me repeat an earlier warning at the risk of redundancy. *Never* concentrate so fully on photography that you forsake *seeing* the eclipse! Even with the finest photographic equipment available, nothing can possibly capture the sheer magic and emotion of an eclipse. The only way to experience such an all-consuming event is to witness it firsthand with the most wondrous optical instrument of all: the human eye.

6 The Best Laid Plans

No matter how much planning, preparation, and foresight you put into viewing an eclipse, something is bound to go wrong. It may be as simple as forgetting to bring along a special tool needed to put a critical piece of equipment together. Or it might be more serious, such as damaging a camera lens or telescope during transit. As the saying goes, however, "forewarned is forearmed." Consider this chapter a forewarning.

THE SITE

First and foremost, plan your trip wisely. Remember the "Rule of the Three L's" so often used in real estate: location, location, location. Before all else, ask yourself, "Where will I be observing the eclipse from?" The answer to that question will influence all subsequent choices, such as equipment selection and the intended observing program.

Ideally, the site should be flat, as high above sea level as possible to avoid low-lying haze, clouds, and fog, and free of obstructions and restrictions that may interfere on eclipse day. It should be readily accessible by roads and other modes of transportation, and be safe from harmful trespassers.

Typically, parks and beaches offer excellent areas. Most enjoy better horizons than other locations, and offer the added benefit of round-the-clock security patrols. Before making plans, however, always ask the park's office if special access is available. Many parks limit their accessibility by residency or to daytime hours only (not a problem for solar eclipses, but a major obstacle to observing lunar ones).

Other good alternatives include schoolyards, campgrounds, and even farmers' fields, but get permission before setting up! Golf courses almost always have wide open expanses, but may also suffer from restrictions. (Golfers trying to play through might also have some thoughts about your being there!) If the owner of the course is apprehensive at first, why not offer to run a free observing session for club members, in return for access? Most people will jump at the chance to see an eclipse. Even a roadside rest area might be suitable, as one was for two friends and me during a madcap race against the clouds for the August 1989 total lunar eclipse.

Consider contacting a local astronomy club to ask what their plans for the eclipse are, and if you might join them. This is especially helpful if you will be observing far from home and don't know the area. Not only will they have the inside track on the best viewing sites, but it's also more fun to view the drama of an eclipse with a group. *Sky & Telescope* magazine publishes lists of astronomy clubs in North America and Europe. An international listing is available on the World Wide Web (WWW) from the Astronomical Society of the Pacific at http://maxwell.sfsu.edu/asp/amateur.html. A second, independent listing is available on the Internet at http://www.rahul.resource/regular/clubs-etc/clubsetc.html.

Cloud cover may be affected by the local terrain, especially during solar eclipses. Sometimes it will cause clouds to dissipate; at other times it will actually cause clouds to form. The locale can have a great impact on which will happen. One of the worst places to view an eclipse is where the winds are blowing uphill. The lee side of a range of hills is much safer.

Here's what happened to me during the 1991 July 11 total eclipse I hoped to view from Baja California, Mexico. The tour leader misjudged where we should set up, choosing a site within the influence of the local mountain range. As the temperature dropped, clouds formed in this otherwise parched environment and, in the process, cut a 6-minute, 58-second total phase by over three minutes!

Always check the latest predictions for the eclipse path before committing yourself to a site. Just ask Ron Woodland of Worthington, Massachusetts. In March 1970, a total solar eclipse track passed very nearly parallel to the East Coast of North America. It left land around Virginia Beach, Virginia, touching Nantucket and Monomoy Island off the Massachusetts coast, before making its way to Nova Scotia. Woodland chose to view the event from Monomoy National Wildlife Refuge, on the extreme northern limit of totality. From there it was predicted that totality would last just 20 seconds—a "graze" eclipse. Little did anyone know, but the predicted track was slightly off, causing him instead to witness a mere *five seconds* of totality. Still, Woodland considered himself lucky after he looked across the open field and saw another group of observers just a quarter-mile away, who missed totality entirely!

THE TRAVELING TELESCOPE

In this age of ever-increasing light pollution, most amateur astronomers find it necessary to travel to remote observing sites just to catch a glimpse of an unspoiled night sky. Therefore, traveling with a telescope to view an eclipse may not be a totally unfamiliar idea to many readers. If so, fine—travel to an eclipse in exactly the same way you would travel to a favorite dark-sky site, if possible. Of course, it's one thing to drive a short distance, such as to a local schoolyard, to show the wonder of an eclipse to a group of children; it's quite another to travel hundreds, even thousands, of miles from home to capture a special perspective of the event.

Before embarking on a cross-country trek, or even a cross-state one, first stop and think about what is entailed when moving a delicate piece of optical equipment. Of course, transporting a large telescope is much more difficult than moving, say, a 400-mm telephoto lens, but regardless of size, the possibility of damage is always present.

Let's first examine traveling by car. Are your car and telescope compatibly sized? If you own a subcompact car and a "full-size" telescope, you might have to make alternate arrangements. Assuming that both can fit together comfortably, then it is simply a matter of storing the instrument safely for transport. First, seal the optics from possible dust and dirt contamination. This usually means simply leaving the dust caps in place, just as when the telescope is stored at home. Take apart as little of the telescope as necessary; unless the instrument is especially small, however, it is likely that the telescope tube will have to be separated from its mounting. Place the optical tube assembly into the car first. If the telescope did not come with a storage case, protect the tube from bumps by wrapping it in a clean blanket, quilt, or sleeping bag. Strategically placed pillows and pieces of foam rubber can also help minimize screw-loosening vibrations. If you intend to place it on the front or back seat, strap it in place using the car's seat belt. You don't want to crash-test your telescope!

Next comes the mounting. Carefully place it into the car, making sure that it does not rub against anything that may damage it, or that it may damage. Again, it is best to wrap everything with a clean blanket for added protection. Consider using either the car's seat belts or bungee cords to keep things from moving around during sharp turns. Be sure to secure counterweights and other heavy, loose items that might become airborne if you have to hit the brakes hard.

Transporting telescopes by any form of public transportation (air, rail, or bus) presents many additional problems. The large dimensions of most telescopes usually make it impossible to carry them on board and store them easily. Therefore, great care must be taken when packing these delicate instruments.

Surprisingly, there is no industry-wide policy for transporting a telescope on a plane. Some airlines permit telescopes to be checked as luggage, provided they do not exceed size and weight restrictions. Others will not accept telescopes as check-in items at all. In those instances you must ship the instrument separately ahead of time via an air-cargo carrier. Because of the amount of paperwork involved, especially on international shipments, air-cargo services usually require that advance arrangements be made. In any case, you are best advised to contact your carrier's customer service office long before your departure date.

Some owners of 8-inch and smaller Schmidt-Cassegrain telescopes prefer to sheath their instruments in foam rubber and place them in large canvas duffel bags. They then carry them on board, placing the instruments into the overhead compartments. Though this method may work in some cases, I can neither endorse nor recommend it. The chance of damaging the telescope is far too great, and there is no way to guarantee that all overhead compartments will be large enough for the instrument to fit!

Most telescope carrying cases are not designed for travel as checked luggage. When I asked a representative from Meade Instruments how they carry their telescopes to trade shows and other events, he recommended constructing a wooden crate from two-by-two (50 mm × 50 mm) wooden framing and ½-inch (12 mm) plywood side panels. The interior dimensions should be the same as the manufacturer's case, so that the same precut foam padding can be used to line the interior. Complete the crate with a pair of strong hinges and a lock.

Celestron says that their *hard plastic* cases, such as those that come with their top-of-the-line models, are strong enough to take the jostling that air cargo can go through. These cases can also be used with other catadioptric instruments, although they may not provide a custom fit.

Even with a high-impact case, take a few additional precautions before sending your "baby" to the loading ramp. Begin by completely enclosing the telescope in protective bubble wrap, available from larger post offices, stationery stores, and shipping firms. Another measure of protection is to place the telescope (case and all) into a heavy-duty, multi-walled cardboard box. These boxes may also be purchased from independent companies that specialize in wrapping packages for shipment.

Owners of refractors and reflectors (especially large-aperture instruments) face even greater challenges. First and foremost, the optics must be removed and placed into a hard case to be carried on by hand. The empty tube may then be bubble-wrapped, surrounded by Styrofoam "popcorn" (both inside and outside the tube) and placed into a strong wooden crate. Make certain that the shipping carton can be used again on the return trip, and bring along a roll of packing tape or duct tape just in case an emergency repair is needed.

Because of their weight, tripods and mountings pose special problems. I have traveled by air with a large tripod by first wrapping it in two thick sleeping bags and then packing everything in a large tent carrying case. Though this may work for most camera tripods, heavy equatorial mountings are best packed professionally. Once again, seek a local crating company for assistance.

Make sure that each piece of luggage has both a destination ticket and an identification tag, and that both are clearly visible on the outside. Information on the ID tag should include your name, complete address, and telephone number. Although permanent, plastic-faced identification tags are preferred, check-in counters provide paper tags that may be filled out on the spot. I always make it a habit to include a second identification label inside my luggage as well, just in case the outside tag is torn off.

Whatever you do, be certain to bring along all the tools needed to reassemble the equipment once you arrive. Remember everything—all the screwdrivers, wrenches, and any special tools. Better yet, bring two of each critical item, plus some spare nuts and bolts, just in case. Pack everything into luggage that will be checked; never carry tools on board, because airport security is likely to view them as potential weapons. While returning from the July 1991 total solar eclipse in Mexico, a friend was stopped from boarding his flight because he was carrying a small screwdriver in his pocket. Although my friend is certainly not prone to violence, he ultimately had to check his "weapon." Unfortunately, all of his luggage had already been loaded on the plane, so the screwdriver had to be checked separately. It was quite a sight in the baggage-claim area when his lone screwdriver came down the ramp in an open-topped plastic box among all of those suitcases!

THE INTERNATIONAL SCENE

For die-hard solar-eclipse fanatics, traveling to exotic foreign lands is par for the course. But even well-planned journeys are usually not without their moments. Almost everyone who has ever traveled has experienced minor inconveniences like losing luggage. Be sure to read everything and know the requirements of each country you are visiting *before* leaving home.

All countries place different restrictions on visitors traveling inside their borders. It goes without saying that you should check that your passport is still valid. Also find out if there are any special restrictions on carrying telescopes, binoculars, and cameras. Generally, items for personal use can be taken in free of customs duties, but travelers would be well advised to check before leaving home. Purchase a reliable travel handbook (such as one of the Fodor or Birnbaum series), and read it from cover to cover. Learn all there is to know about travel to and within your country of destination. Further, for readers with access to the Internet, connect directly with the United States Department of State for the latest travel requirements and advisories. Their electronic address is http://www.stolaf.edu/network/travel-advisories.html.

Visas may also be required to enter a foreign country. A visa is stamped onto your passport by an official from that country to show that you have been approved for travel. To find out if the country (or countries) you will be visiting requires a visa, check with its state department or travel authority. Your travel agent or tour organization should also be able to provide the latest information.

Residents of the United States should compile a thorough inventory of all equipment they are taking out of the United States. U.S. Customs requires owners to register cameras and accessories on a "Certificate of Registration for Personal Effects Taken Abroad" form before departure. This form has space for a complete description of each item, such as dimensions, color, serial number, manufacturer, and approximate value. Be sure to register any homemade telescopes and other equipment as well, marking them with some sort of "dummy" serial number for identification. Veteran solar-eclipse traveler Ernie Piini found this out the hard way when he was detained after landing in Uruguay for the 1992 June 30 eclipse because of an unregistered homemade telescope. Not only will this inventory prove invaluable when passing through customs, but it is also a worthwhile document when trying to locate and identify a lost or stolen piece of equipment. Contact your nearest U.S. Customs office for further information.

Keep in mind that, if you travel by air, both you and your luggage will have to pass through X-ray scanners at least a couple of times. Because those scanners may fog film, especially those in excess of about ISO 200, it is best to ask the official at the airport entrance to inspect your camera bag visually. Placing all rolls of film into a clear plastic bag will greatly speed things along.

Be aware that photography in certain "politically active" regions, such as Eastern Europe and parts of the Middle East, can be a sensitive subject. In those areas, do not photograph government institutions or personnel, military bases, airports, or police. Even giving the appearance of taking a photograph may lead to your film, and perhaps even your camera, being confiscated by authorities. If in doubt about what is safe and unsafe to photograph, always ask first.

In those same countries, it is advisable to register with your country's nearest embassy or consulate. Representatives are there to help in case there is an emergency or an accident, or even if you have lost your passport. Before leaving home, find out where each embassy or consulate is located along your journey.

Health issues are always a topic of consideration when traveling to foreign lands. If you are going with a tour group, be sure to check with the tour's leader for specific requirements. If you are traveling on your own or with a small, informal party, contact the Centers for Disease Control and Prevention in Atlanta. By calling (404) 332-4559, you will reach a 24-hour hot line that lists the latest international health requirements and advisories for foreign travelers. If you live outside the United States, contact your local health office for further recommendations.

Should you have a chronic or acute medical condition, it goes without saying that you should see your doctor before traveling. Also make sure that your medical insurance is valid outside of the United States, and that you have ready access to cash, should an emergency crop up while you are away from home. For a directory of English-speaking doctors abroad, contact the International Association of Medical Assistance, at 417 Center Street, Lewiston, NY 14092; in Canada at 40 Regal Street, Guelph, Ontario N1K 1B5; and in Europe at 57 Voirets, 1212 Grand-Lancy-Geneva, Switzerland.

Did you get your shots? Not all countries require inoculations before visitation, but many do. At least two weeks before departure, check with your personal physician or the Centers for Disease Control and Prevention for up-to-date information on the occurrences of communicable diseases and advice on which, if any, inoculations are recommended.

Finally, many countries supply different voltages from that used in the United States. For instance, the electrical services in most European countries are based on 200–250 volts, 50 cycles, which is very different from the 110-to-120-volt, 60-cycle current used in the United States and Canada. This difference will cause electrical equipment to malfunction and become damaged. If you are bringing along a motor-driven telescope mount or other article that must be plugged into a wall socket, make sure you get a proper transformer (available at most electrical-supply stores) to convert between the voltages.

IS GETTING THERE *REALLY* HALF THE FUN?

Now there's a question that is often pondered by eclipse chasers. While lunar eclipses are usually viewed from close to home, total and annular solar eclipses can lead to driving along unpaved roads in dilapidated buses or flying high above the ground in a plane that seems barely able to stay aloft. Yes, getting to the eclipse can be *almost* as exciting as the event itself at times.

This leads to a second question: When traveling to see an eclipse, is it better to go with an organized tour, or to travel independently? Since the 1970s, solar-eclipse tours sponsored by professional tour agencies, larger astronomy clubs, and museums have become increasingly popular (as evidenced by Appendix D). They offer a certain sense of security that traveling alone does not. Larger tours will have already arranged for an observing site as well as transportation to and from the hotel, and will have likely hired a local tour guide who knows all of the region's ins and outs. Security will also be provided, should the area require it.

Oceangoing tours on board luxury liners enable observers to see eclipses that occur entirely over open water (figure 6.1), while land-based tours often book the best accommodations long before individuals are able to do so. Both will frequently include side tours to various attractions of historic, astronomical, or general interest. Many prearranged tours will also offer seminars and lecture series on photography, astronomy, nature, and many other areas of interest by leading authorities.

Traveling on your own, however, does have several distinct advantages. First and foremost is the potential for monetary savings. Rather than staying at preselected hotels for a dictated number of days, the independent traveler can create an itinerary to suit his or her own tastes. This means that if clouds or inclement weather seem to be a foregone conclusion on eclipse day, the independent traveler can easily pull up stakes and perhaps locate some clear skies.

Nevertheless, with independence comes headaches. If you're going it alone or leading a small group, it is absolutely critical to make accommodation and transportation reservations as far ahead of time as possible, to ensure their availability. Bear in mind that these usually require advance payment. Even then, some hotels may be reluctant to book you and your party, instead trying to hold out for bigger tour groups. If this is the case, try contacting smaller hotels and inns. Frequently bypassed by large groups, these can offer equal, if not superior, accommodations to those offered by their larger counterparts. And of course there's always the possibility of roughing it in a campground. If this latter approach is your preference, again be sure to check any restrictions and safety advisories.

Figure 6.1 Many observers prefer the luxury and mobility of solar-eclipse chasing from on board an ocean liner. This photograph was taken on the SS *Jubilee* by Spencer Rackley IV during the July 1991 solar eclipse.

If you are traveling to a distant location to watch an eclipse, plan on arriving several days beforehand. Much of that time will be spent scouting out a suitable viewing site, following the advice outlined earlier.

The next chapter gives some statistical data on climatological conditions for each eclipse, but it is best to contact a local radio meteorologist or weather office to get the latest forecasts. If connection to the Internet is available, it might also be possible to view the latest satellite images directly up to the moment when a decision has to be made if conditions are worsening. In this way, you and your group can head off in the right direction in an attempt to beat the clouds. Four of the best starting points are the weather pages at INTELLiCast (http://www.intellicast.com/weather), the Weather Channel (http://www.weather.com), the University of Michigan (http://cirrus.sprl.umich.edu), and the Goddard Space Flight Center (http://climate.gsfc.nasa.gov).

Another way to obtain the latest data, even for the most remote corners of the Earth, is to link up directly with an orbiting weather satellite. Any single-sideband shortwave radio can receive direct facsimile transmissions from weather satellites, which are then fed into a personal computer using a special radio-to-computer interface cable. The computer demodulates the signal using special imaging software to produce up-to-the-minute views. One company that sells signal demodulators and the requisite software is Software Systems Consulting (see Appendix A).

Two very important words when planning to travel to a solar eclipse are these: *Be creative.* Ponder all possibilities, and never dismiss an option that might seem ridiculous at first. To illustrate this, consider the improvised eclipse expedition put together for the July 1990 total eclipse by Joe and Renate Rao, Sam Storch, and Craig Small, four amateurs from New York and New Jersey. Studying commercial airline flight paths and schedules, they found that a regularly scheduled American Trans-Air flight between Hawaii and San Francisco would pass very close to the path of total-

Figure 6.2 Some solar-eclipse chasers go to any length to bask in the shadow of the Moon! Note the cone-shaped umbral shadow. This photograph of the July 1990 total solar eclipse was taken from on board a regularly scheduled American Trans-Air flight between Hawaii and San Francisco. Photo by Craig Small. (24-mm lens at f/2.8, metered (automatic) exposure, and Kodachrome 64 film.)

ity. All that was needed was a slight adjustment to the flight plan, delaying the departure for approximately 40 minutes. Airline officials agreed to their request, and these intrepid four, plus the rest of the passengers and crew, flew within totality for 73 seconds (figure 6.2)!

By land, by sea, or by air, some will go to great lengths to see an eclipse. To use an old cliché, where there's a will, there's a way.

7 Solar Eclipses: 1998–2017

I'd like to expand on Benjamin Franklin's time-honored maxim: In this world nothing is certain but death and taxes—and *eclipses!* And what does the future hold? Gazing into the eclipse crystal ball reveals a wide variety of events, both solar and lunar. This chapter takes a close-up look at the next twenty years of solar eclipses, while lunar eclipses are discussed in chapter 8.

Both chapters list upcoming eclipses in chronological order. In this chapter, each event features a general discussion of where the eclipse will be visible, with the central track of each total, annular, and annular-total eclipse plotted on an accompanying map. In addition, for readers who may wish to plot the course of an eclipse on a detailed world map for themselves, each total and annular event includes a table listing exact latitude and longitude coordinates. Note that in the tables to come, north latitudes are shown by a plus sign (as in +40°), while south latitudes are denoted with a minus sign. Likewise, west longitudes are noted as a negative value (e.g., to specify 73° west longitude, the tables would show −73°), while east longitudes are shown as positive values.

Most of the discussions also include predicted times of the various contacts. These contacts are defined in table 7.1. All times have been rounded down to the nearest whole minute and are specified in Universal Time, or U.T. (see Appendix E for an explanation).

Durations of all total and annular eclipses are also itemized along their tracks. Each eclipse's *instant of greatest eclipse* is annotated with "GE" listed parenthetically next to the corresponding set of coordinates. Solar-eclipse expert Fred Espenak of the Goddard Space Flight Center in Greenbelt, Maryland, explains just what "greatest eclipse" means.

Table 7.1 **Eclipse Contact Abbreviations**

P1	first penumbral contact (beginning of partial eclipse)
U1	first umbral contact (beginning of totality or annularity)
Maximum	time of maximum eclipse
U2	second umbral contact (end of totality or annularity)
P2	second penumbral contact (end of partial eclipse)

"Greatest eclipse" is defined as the instant when the center of the Moon's shadow passes nearest the Earth's geocenter. It is *approximately* equal to the moment of "greatest magnitude" and "greatest duration" for total solar eclipses, but does not match exactly because of the tilt of Earth's axis, tilt of Moon's orbit, latitude of the shadow's path, and oblateness of Earth.

For annular solar eclipses you have two effects that conspire against each other to move the instant of "greatest duration" significantly away from "greatest eclipse." As the shadow approaches the position of "greatest eclipse," its ground velocity slows down, resulting in a longer eclipse. At the same time, however, Earth's curvature brings the observer deeper into the Moon's shadow, causing the Moon's apparent diameter to increase. This results in a shorter annular eclipse. It's the balance of these two effects that seems to move the point of "greatest duration" around so much.

Additional tables list the weather prospects for selected cities in the path of the eclipse. No, I'm not trying to play meteorologist; rather, I offer only a brief numerical synopsis based on statistics from the National Climatic Data Center's *International Station Meteorological Climate Summary* database.[1] The database contains climatological measurements pooled from decades of weather observations made at United States Air Force airfields, commercial airports, and other institutions around the world.

Rather than specifying exhaustive facts about each of the climate-observation points cited here, I have restricted each listing to sky conditions around the time when mid-eclipse will occur from that location. In most cases the data is broken down into four categories according to sky obscuration: *clear* corresponds to no cloud cover (i.e., the sky is perfectly clear); *scattered* means that between 10 percent and 50 percent of the sky is hidden by clouds; *broken* indicates that between 60 and 90 percent of the sky is covered, while *overcast* means that the sky is completely obscured. (The percentages should tally to 100 percent; any discrepancy is due to unreported data.) Clearly, the best prospects for seeing an eclipse will be at locations with high *clear* and *scattered* values, and low *broken* and *overcast* percentages.

Unfortunately, not all worldwide observation stations record data in this manner. As a result, you will come upon some entries that infer sky cover in terms of two criteria: "the number of days within the given month when cloud cover was less than 30 percent" and "the number of days with more than 0.1 inch of precipitation." The former appears abbreviated in the tables ahead as "Clouds < ³⁄₁₀," while the latter has been abbreviated "Precip > 0.1 inch," each being followed by a numerical value that represents the number of days within the eclipse's month when that criterion was met. Areas with relatively cloud-free track records will have high "Clouds < ³⁄₁₀" values and low "Precip > 0.1 inch" values.

Although each set of eclipse predictions listed here and in chapter 8 has been checked and cross-checked using accurate data and algorithms, subtle nuances and refinements are likely to occur in subsequent years. Therefore I *strongly* recommend that, as each eclipse draws closer, you consult astronomical periodicals, such as *Sky & Telescope* or *Astronomy*, or annual publications such as Guy Ottwell's *Astronomical Calendar* or the Royal Astronomical Society of Canada's *Observer's Handbook*, for the latest predicted times and contact points. In addition, NASA's resident eclipse expert, Fred Espenak, and Jay Anderson, a meteorologist with Environment Canada, coauthor bulletins predicting solar-eclipse tracks two years before the featured event. As of 1997, they were available via the Internet's World Wide Web at http://umbra.gsfc.nasa.gov/eclipse/predictions/eclipse-paths .html, or from the address in Appendix F.

That having been said, here are capsule summaries for the next two decades of solar eclipses.

TOTAL SOLAR ECLIPSE OF 1998 FEBRUARY 26 (SAROS 130)

This is the total eclipse that many of us have been waiting for since the great eclipse of July 1991. Unlike the winter eclipse of 1997, when bone-chilling prospects in Mongolia and Siberia kept many observers tucked warmly at home, this event is sure to attract observers seeking a little quality time in the umbra as well as a midwinter reprieve from the snows back home.

The shadow of the Moon first touches the Earth in the South Pacific Ocean. Racing eastward across the ocean for several thousand miles, totality skims across the northern edge of the Galápagos Islands, 1,200 kilometers (750 miles) off the coast of Ecuador. Cruise ships may well set sail for the centerline to the north of the islands, where totality will last over 4 minutes. Greatest eclipse, with 4 minutes 9 seconds of totality, occurs in the Pacific some 580 kilometers (360 miles) west of the Colombian coast.

The shadow continues its watery trek to the northeast, next touching dry land at the Panama/Colombia border (figure 7.1). It then continues across northern Colombia, passing over Montería, Valledupar, and several smaller towns. Maracaibo and other towns in northern Venezuela are next to greet the onrushing shadow, as it heads back out to sea.

Several islands in the Netherlands Antilles lie within the shadow's path. Of these, many eclipse chasers will set up on Aruba and Curaçao, where totality will last a few seconds over 3 minutes. While both islands will be fully immersed in the shadow of the Moon, the best observing sites will be found along Aruba's south shore near the towns of Sint Nicolaas and Seroe Colorado, and Curaçao's northwest corner near Santa Cruz, Westpoint, and Saint Christoffel National Park, as the proximity of the central line promises slightly longer encounters.

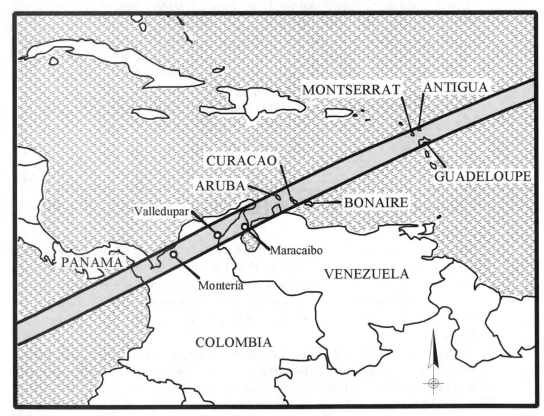

Figure 7.1 Track of the total solar eclipse of 1998 February 26.

Crossing the Caribbean Sea, totality also befalls the Leeward Islands of Antigua, Montserrat, and Guadeloupe before heading into the open Atlantic. The north shore of Guadeloupe's Grand-Terre island, with its capital of Anse-Bertrand and Port-Louis, offers the best view of totality, as it lies closest to the path's center. Montserrat and Antigua lie very close to the northern edge of the track, thereby giving viewers only a fleeting glimpse of totality. The best sites on Montserrat include the area south of Plymouth (the capital of Montserrat), although recent volcanic activity might make this a less attractive observing site. On Antigua, the towns of Falmouth and English Harbor offer the best views.

Crossing the Atlantic in the late afternoon, the shadow finally leaves the Earth just short of the Canary Islands and the African continent.

If you are planning to see this eclipse and have not already made travel arrangements, time is growing short! Although there are several hotels on each of the islands, as well as in South America, most will be completely booked by the time you read this. A better approach at this late date would be to contact the tours listed in Appendix D and ask about their packages. Tour groups also advertise extensively in both *Sky & Telescope* and *Astronomy* magazines; see Appendix B.

For those staying home, a partial eclipse awaits all of Mexico and Central America, as well as a good portion of South America and the United States. Specific times are offered below.

Table 7.2 **Total Solar Eclipse of 1998 February 26**
Center Line Coordinates

U.T. hh:mm	Longitude °	′	Latitude °	′	Alt. °	Path Width km	Duration m:ss
15:47	−140	00	−02	56.7	04	92	1:32
15:51	−130	00	−04	02.9	15	105	1:55
15:59	−120	00	−04	29.3	27	118	2:22
16:12	−110	00	−04	02.9	40	131	2:55
16:34	−100	00	−02	18.7	55	143	3:31
17:03	−90	00	+01	06.7	70	151	4:01
17:29	**−82**	**42.7**	**+04**	**42.9**	**76**	**151**	**4:09 (GE)**
17:38	−80	00	+06	14.1	75	151	4:07
18:10	−70	00	+12	06.9	62	144	3:44
18:35	−60	00	+17	29.9	46	134	3:09
18:52	−50	00	+21	52.4	32	123	2:37
19:02	−40	00	+25	15.6	21	112	2:07
19:07	−30	00	+27	49.1	10	94	1:33
19:09	−20	00	+29	41.6	01	92	1:30

Table 7.3 **Local Contact Times: 1998 February 26**
     a. Total

City/town	P1	U1	Maximum	U2	P2	Duration m:ss
Montería, Colombia	16:17	17:50	17:52	17:54	19:22	3:57
Maracaibo, Venezuela	16:31	18:04	18:05	18:06	19:31	2:54
Sint Nicolaas, Aruba	16:38	18:09	18:11	18:13	19:36	3:30
Oranjestad, Aruba	16:38	18:09	18:11	18:12	19:35	2:49
Willemstad, Curaçao	16:40	18:12	18:13	18:14	19:37	1:57
Santa Cruz, Curaçao	16:40	18:11	18:13	18:14	19:37	3:20
Anse-Bertrand, Guadeloupe	17:05	18:31	18:32	18:34	19:49	3:01
Basse-Terre, Guadeloupe	17:04	18:31	18:31	18:32	19:49	1:20
Falmouth, Antigua	17:05	18:31	18:32	18:33	19:49	2:38
St. John's, Antigua	17:05	18:31	18:32	18:33	19:49	2:07

     b. Partial

City	P1	Maximum	P2	Magnitude
Atlanta	16:59	17:57	18:53	0.251
Bogotá	16:17	17:51	19:19	0.879
Caracas	16:44	18:15	19:38	0.926
Chicago	17:30	17:55	18:20	0.048
Dallas	16:44	17:27	18:11	0.136
Detroit	17:31	18:04	18:36	0.086
Miami	16:43	18:00	19:15	0.496
New York City	17:29	18:19	19:08	0.218
Washington, D.C.	17:22	18:14	19:04	0.224

Table 7.4 **Weather Prospects: 1998 February 26**

City	Clear	Scattered	Broken	Overcast
Basse-Terre, Guadeloupe	0	56.8	42.4	0.8
Beatrix Airport, Aruba	0.2	73.1	26.5	0.2
Coolidge Airport, Antigua	0	60.4	38.5	1.1
Hato Airport, Curaçao	0	68.4	30.9	0.8
Maracaibo, Venezuela	2.9	33.6	38.5	25.0
Montería, Colombia	1.3	30.1	62.7	5.9

ANNULAR SOLAR ECLIPSE OF 1998 AUGUST 22 (SAROS 135)

This annular eclipse is best characterized as an "island-hopper's dream," as the path crosses many of the thousands of islands in Malaysia and Indone-

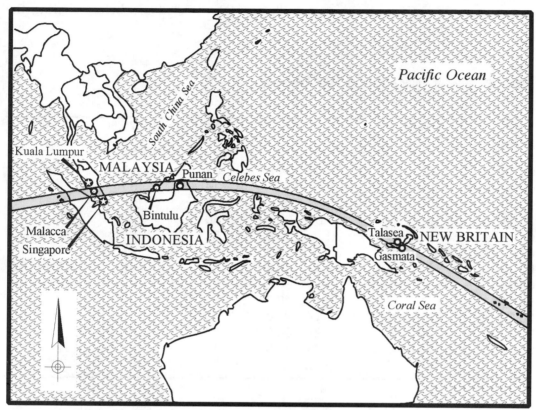

Figure 7.2 Path of the annular solar eclipse of 1998 August 22.

sia, in between mainland Asia and Australia (figure 7.2). The eclipse begins in the Indian Ocean, first striking land on the island of Sumatra and passing over the towns of Padang and Sibolga. There, annularity lasts some 2 minutes 48 seconds, but with the Sun less than 5° above the horizon.

Crossing the Strait of Malacca, the eclipse track hits the southern portion of the Malaysian peninsula between the cities of Kuala Lumpur and Singapore. The towns of Malacca, Bandar, Segamat, and Mersing all witness a few seconds less than 3 minutes of annularity.

Heading into the South China Sea, the shadow next strikes Sarawak, the Malaysian-owned area of the island of Borneo, where 2 minutes 54 seconds of annularity is witnessed in the seaside town of Bintulu. Border-hopping into the Indonesian portion of Borneo, the shadow passes over the town of Punan before moving into the Celebes Sea.

Annularity narrowly misses the north shore of New Guinea. Instead, greatest eclipse occurs some 120 kilometers (75 miles) offshore from Cape Girgir. The path does, however, strike the island of New Britain, where annularity lasts 2 minutes 59 seconds. The shadow then curves farther to the southeast, passing between Fiji and New Caledonia before leaving the Earth's surface.

Table 7.5 **Annular Solar Eclipse of 1998 August 22**
Center Line Coordinates

U.T. hh:mm	Longitude °	′	Latitude °	′	Alt. ′	Path Width km	Duration m:ss
00:15	+90	00	−00	04.8	03	145	2:45
00:19	+100	00	+01	45.1	14	137	2:49
00:28	+110	00	+03	02.1	26	127	2:54
00:43	+120	00	+03	27.2	40	116	3:01
01:08	+130	00	+02	30.6	55	108	3:08
01:43	+140	00	−00	26.2	71	101	3:14
02:07	**+145**	**23.4**	**−02**	**59.3**	**75**	**99**	**3:14 (GE)**
02:25	+150	00	−05	36.9	72	99	3:13
03:03	+160	00	−11	48.7	56	105	3:07
03:28	+170	00	−17	26.7	40	117	3:02
03:44	180	00	−22	00.6	26	129	2:57
03:52	−170	00	−25	35.3	14	141	2:53
03:56	−160	00	−28	21.8	04	151	2:51

Table 7.6 **Local Contact Times: 1998 August 22**
a. Annular

City/town	P1	U1	Maximum	U2	P2	Duration m:ss
Malacca, Malaysia	23:10	00:19	00:20	00:22	01:44	2:48
Bintulu, Malaysia	23:13	00:30	00:31	00:33	02:08	2:54
Punan, Indonesia	23:15	00:35	00:36	00:38	02:17	2:58
Talasea, New Britain	00:27	02:24	02:25	02:26	04:15	2:26

b. Partial

City	P1	Maximum	P2	Magnitude
Auckland, New Zealand	02:20	03:37	04:46	0.525
Brisbane, Australia	01:26	03:00	04:25	0.493
Darwin, Australia	23:54	01:31	03:20	0.586
Melbourne	01:46	02.45	03:43	0.175
Singapore	23:10	00:22	01:47	0.954
Sydney	01:41	02:59	04:11	0.325

Table 7.7 **Weather Prospects: 1998 August 22**

City	Clear	Scattered	Broken	Overcast
Bintulu, Malaysia	0	4.2	84.5	11.3
Malacca, Malaysia	0	0.3	90.4	9.3

ANNULAR SOLAR ECLIPSE OF 1999 FEBRUARY 16 (SAROS 140)

A summer eclipse always promises a pleasant activity surrounded by all that nature has to offer. That's just what this late-afternoon annular event guarantees viewers who are set along a narrow west-to-east line that spans Australia (figure 7.3). After traveling across much of the southern Indian

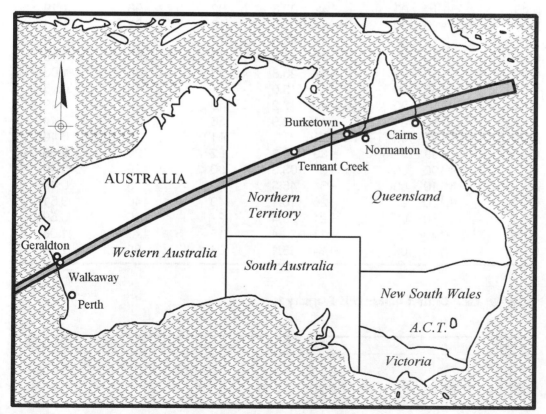

Figure 7.3 Path of the annular solar eclipse of 1999 February 16.

Ocean, the Moon's shadow first touches land on the Australian west coast. There, the towns of Walkaway and Greenough witness a short 35-second annular phase.

Tracking northeastward, the shadow crosses the desert region of interior Australia. The small town of Tennant Creek enjoys 62 seconds of annularity, while farther east, annularity lasts for 65 seconds from the Queensland town of Burketown. By this time, however, the Sun is low in the northwestern sky, setting shortly after fourth contact.

All of Australia will be treated to a partial eclipse sometime during the day, as will a portion of South Africa. Exact times and magnitudes are listed below.

Table 7.8 **Annular Solar Eclipse of 1999 February 16**
Center Line Coordinates

U.T. hh:mm	Longitude °	′	Latitude °	′	Alt. °	Path Width km	Duration m:ss
04:57	+10	00	−42	00.6	02	96	1:19
04:59	+20	00	−44	02.5	10	82	1:14
05:03	+30	00	−45	40.0	17	73	1:10
05:09	+40	00	−46	47.9	25	63	1:05
05:18	+50	00	−47	20.8	33	54	1:00
05:29	+60	00	−47	13.0	41	45	0:55
05:44	+70	00	−46	17.2	49	38	0:49
06:02	+80	00	−44	23.9	56	32	0:44
06:24	+90	00	−41	21.1	61	29	0:40
06:34	**+93**	**53.6**	**−39**	**49.6**	**62**	**29**	**0:39 (GE)**
06:49	+100	00	−37	03.3	60	29	0:40
07:16	+110	00	−31	45.3	52	35	0:43
07:39	+120	00	−26	16.1	40	46	0:51
07:55	+130	00	−21	25.4	27	60	0:59
08:05	+140	00	−17	32.4	15	73	1:06
08:09	+150	00	−14	35.1	04	85	1:11

Table 7.9 **Local Contact Times: 1999 February 16**
a. Annular

City/town	P1	U1	Maximum	U2	P2	Duration m:ss
Walkaway, Australia	06:00	07:27	07:28	07:28	08:44	0:36
Tennant Creek, Australia	06:48	08:00	08:00	08:01	09:04	1:02
Burketown, Australia	06:56	08.04	08.05	08:05	09:06	1:05

Table 7.9 **(Continued)**

b. Partial

City	P1	Maximum	P2	Magnitude
Adelaide, Australia	06:33	07:39	08:39	0.632
Brisbane, Australia	06:54	07:53	—	0.674
Canberra, Australia	06:43	07:41	08:35	0.548
Cape Town	—	04:49	05:50	0.725
Darwin, Australia	06:57	08:08	09:11	0.784
Melbourne	06:36	07:36	08:31	0.531
Perth, Australia	05:59	07:24	08:40	0.919
Sydney	06:46	07:44	08:37	0.561

Table 7.10 **Weather Prospects: 1999 February 16**

City	Clear	Scattered	Broken	Overcast
Geraldton, Australia	Clouds <³⁄₁₀ = 16.4 days		Precip >0.1 inch = 1.8 days	
Normanton, Australia	4.2	44.6	25.3	4.1
Tennant Creek, Australia	3.4	28.8	59.6	8.2

TOTAL SOLAR ECLIPSE OF 1999 AUGUST 11 (SAROS 145)

This is the eclipse that central Europe has been waiting more than three decades for! Not since 1961 February 15 has the continent been treated to such a spectacular total solar eclipse.

The Moon's umbral shadow first touches down off the south shore of Newfoundland in Canada, where early risers across the island province are treated to a dramatic view of the Sun rising in deep eclipse. More than 90 percent of the solar disk is covered from the city of St. John's, but a good view to the east-northeast is needed, since maximum eclipse occurs when the Sun is only some 5° above the horizon.

Crossing the North Atlantic, the umbra makes landfall on the southern shore of Great Britain, where its northern edge glances across Plymouth and several surrounding towns (figure 7.4a). Crossing the English Channel, totality scurries over northern France. While totality narrowly misses Paris, several smaller towns and cities, including Le Havre, Rouen, and Strasbourg, lie in its path.

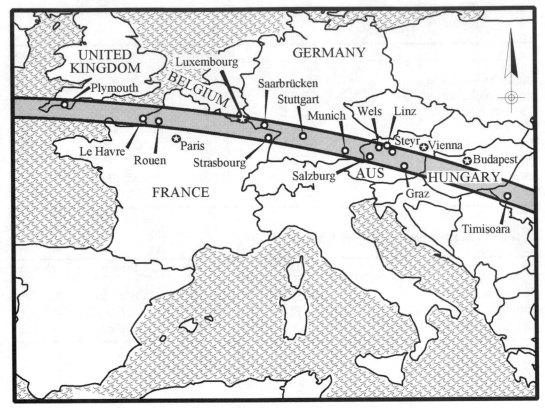

Figure 7.4a Track of the total solar eclipse of 1999 August 11.

Farther east, the shadow immigrates briefly into Luxembourg before crossing into Germany. There the track passes over many towns and cities in the southern third of the country, including Stuttgart and Munich, each of which sees over two minutes of totality.

The length of totality increases as the shadow crosses Austria and Hungary, where it passes just south of their respective capitals of Vienna and Budapest, before pushing into Romania (figure 7.4b). Greatest eclipse, with 2 minutes 23 seconds of totality, occurs a few kilometers southwest of the Romanian town of Rîmnicu-Vîlcea. Though this village is comparatively remote, on the southern slopes of the Transylvanian Alps, Romania's capital of Bucharest is treated to 2 minutes 22 seconds of totality.

The northeastern corner of Bulgaria hosts the eclipse just before it sets sail across the Black Sea and into Turkey. Once back on land, the shadow covers small portions of Syria and Iraq and cuts a northwest-southeast diagonal course across Iran. Finally the path of totality traverses the coast of Pakistan before making its way into India (figure 7.4c), where the shadow leaves the Earth near the towns of Nāgpur and Hyderābād.

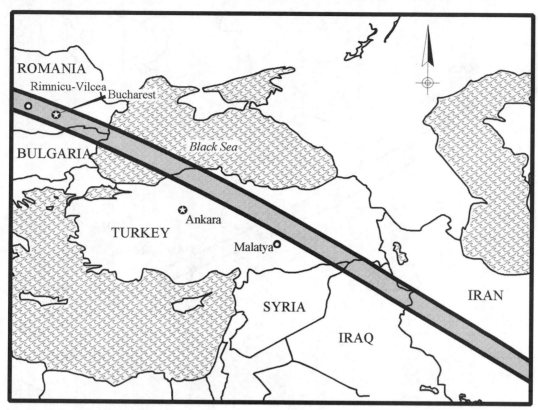

Figure 7.4b Track of the total solar eclipse of 1999 August 11.

Table 7.11 **Total Solar Eclipse of 1999 August 11**
Center Line Coordinates

U.T. hh:mm	Longitude °	′	Latitude °	′	Alt. °	Path Width km	Duration m:ss
09:30	−60	00	+42	26.0	04	61	0:49
09:33	−50	00	+44	59.0	12	75	1:04
9:38	−40	00	+47	07.8	20	83	1:17
09:45	−30	00	+48	46.4	28	91	1:30
09:54	−20	00	+49	49.5	35	97	1:43
10:05	−10	00	+50	12.4	43	102	1:56
10:19	+00	00	+49	50.3	49	106	2:08
10:35	+10	00	+48	36.5	55	109	2:17
10:54	+20	00	+46	22.3	59	111	2:23
11:03	**+24**	**17.7**	**+45**	**04.3**	**59**	**112**	**2:23 (GE)**
11:15	+30	00	+42	59.4	58	112	2:21
11:38	+40	00	+38	28.9	53	111	2:10
12:00	+50	00	+33	15.6	42	105	1:52

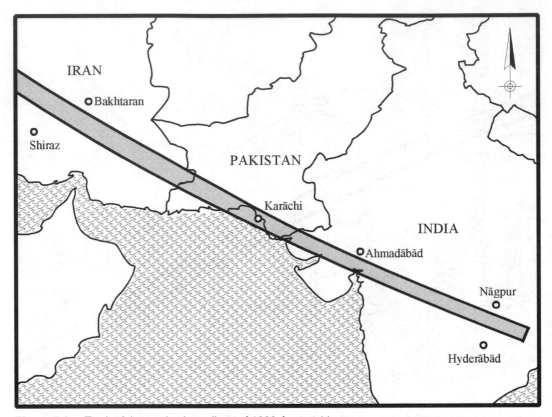

Figure 7.4c Track of the total solar eclipse of 1999 August 11.

Table 7.11 **(Continued)**

U.T. hh:mm	Longitude °	'	Latitude °	'	Alt. °	Path Width km	Duration m:ss
12:18	+60	00	+28	04.1	31	94	1:29
12:29	+70	00	+23	30.0	19	80	1:09
12:34	+80	00	+19	45.6	07	65	0:52

Table 7.12 **Local Contact Times: 1999 August 11**

a. Total

City/town	P1	U1	Maximum	U2	P2	Duration m:ss
Plymouth, England	08:58	10:12	10:13	10:14	11:33	1:32
Le Havre, France	09:02	10:18	10:19	10:20	11:41	1:38
Rouen, France	09:03	10:20	10:21	10:21	11:42	1:34

Table 7.12 **(Continued)**

City/town	P1	U1	Maximum	U2	P2	Duration m:ss
Luxembourg, Luxembourg	09:09	10:28	10:28	10:29	11:51	1:18
Saarbrücken, Germany	09:10	10:29	10:30	10:31	11:52	2:10
Strasbourg, France	09:11	10:30	10:31	10:32	11:54	1:33
Stuttgart, Germany	09:13	10:33	10:34	10:35	11:56	2:16
Munich	09:16	10:37	10:38	10:39	12:01	2:09
Salzburg, Austria	09:18	10:39	10:40	10:42	12:04	2:04
Wels, Austria	09:20	10:41	10:42	10:43	12:05	1:44
Steyr, Austria	09:20	10:42	10:43	10:44	12:06	1:47
Graz, Austria	09:22	10:44	10:45	10:46	12:08	1:19
Timisoara, Romania	09:32	10:55	10:56	10:57	12:19	2:05
Rîmnicu-Vîlcea, Romania	09:37	11:01	11:03	11:04	12:25	2:22
Bucharest, Romania	09:41	11:05	11:07	11:08	12:28	2:22
Karachi, Pakistan	11:18	12:25	12:26	12:27	13:27	1:12

b. Partial

City	P1	Maximum	P2	Magnitude
Istanbul	09:49	11:16	12:38	0.956
London	09:03	10:20	11:40	0.967
Madrid	08:52	10:09	11:33	0.729
Moscow	09:58	11:09	12:18	0.664
Paris	09:04	10:22	11:45	0.992
Rome	09:17	10:42	12:09	0.838
St. John's, Newfoundland	08:39	09:36	10:37	0.929

Table 7.13 **Weather Prospects: 1999 August 11**

City	Clear	Scattered	Broken	Overcast
Bucharest, Romania	39.4	26.9	24.5	9.2
Karachi, Pakistan	4.3	19.5	42.5	33.7
Linz, Austria	12.2	35.0	37.3	15.5
Luxembourg, Luxembourg	13.4	31.0	46.5	9.0
Malatya, Turkey	68.4	29.4	2.2	0
Munich	13.5	36.0	32.7	17.6
Plymouth, England	5.7	25.0	49.9	19.0
Saarbrücken, Germany	12.8	29.5	41.3	15.8
Strasbourg, France	14.0	32.2	40.5	13.3
Stuttgart, Germany	13.3	33.0	40.8	12.9

PARTIAL SOLAR ECLIPSE OF 2000 FEBRUARY 5 (SAROS 150)

The last year of the second millennium hosts four solar eclipses. All are comparatively shallow partial events, likely to draw little attention except in the remote regions where they are immediately visible. Indeed, this first eclipse seems likely to pass unnoticed, as it will only be visible from the Kerguelen Islands (annexed by France) or Antarctica!

Table 7.14 **Local Contact Times: 2000 February 5**
(Greatest eclipse: Time 12:50 u.t., Latitude 70° 13.1'S, Longitude 134° 10.5'W, Mag. 0.579)

Location	P1	Maximum	P2	Magnitude
Scott Base, Antarctica	11:35	12:29	13:23	0.551
Byrd Station, Antarctica	11:15	12:08	13:02	0.444
Port Lockroy, Antarctica	11:06	11:40	12:16	0.123
Kerguelen Islands	13:00	13:49	14:35	0.331

PARTIAL SOLAR ECLIPSE OF 2000 JULY 1 (SAROS 117)

Just as the last eclipse was nearly inaccessible, so too does it seem likely that few will witness this partial eclipse. The eclipsed area is centered over the far southeastern corner of the Pacific Ocean, with the southernmost tip of South America the only land being affected. That day, portions of southern Chile and Argentina will witness a small notch carved from the Sun as sunset nears.

Table 7.15 **Local Contact Times: 2000 July 1**
(Greatest eclipse: Time 19:33 u.t. Latitude 66° 55.9'S, Longitude 109° 27.5'E, Mag. 0.477)

Town/Location	P1	Maximum	P2	Magnitude
Camarones, Argentina	20:04	20:28	20:51	0.073
Puerto Aisén, Chile	19:47	20:22	20:56	0.148
Wellington Island, Chile	19:33	20:16	20:57	0.229

PARTIAL SOLAR ECLIPSE OF 2000 JULY 31 (SAROS 155)

You are probably familiar with the phrase "once in a blue moon," indicating that something occurs only rarely. "Blue moon" refers to the second Full Moon in a given month. Since Full Moons occur once every 29.5 days, most months have only one. Every couple of years on average, the phases will line up such that two Full Moons squeeze into one calendar month. Well, with the coming of this second partial eclipse in July 2000, we can say "once in a blue eclipse"! The last month to feature two solar eclipses was way back in 1880 (December 2 and December 31), with the next not occurring until 2206 (December 1 and December 30). Now, *that's* rare!

This time the Moon's penumbral shadow brushes western Canada, the states in the Pacific Northwest, Alaska, much of Greenland, and the northern third of Russia. Observers in the United States and Canada should note that the eclipse will be visible locally in the late afternoon or early evening on July 30. Greenland will have the unique opportunity to see an eclipse of the midnight Sun, because the event occurs late at night from that vantage point.

Table 7.16 **Local Contact Times: 2000 July 31**
(Greatest eclipse: Time 02:14 u.t., Latitude 69° 31.7′ N, Longitude 59° 50.8′W, Mag. 0.603)

City	P1	Maximum	P2	Magnitude
Anchorage	02:07	02:49	02:29	0.303
Edmonton, Alberta	02:17	02:57	03:37	0.424
Juneau	02:13	02:55	03:36	0.365
Regina, Saskatchewan	02:18	02:57	03:35	0.426
Salem, Oregon	02:39	03:13	03:46	0.268
Seattle	02:33	03:09	03:45	0.309
Vancouver, British Columbia	02:30	03:07	03:44	0.327

PARTIAL SOLAR ECLIPSE OF 2000 DECEMBER 25 (SAROS 122)

The final eclipse of the twentieth century comes wrapped as a Christmas present for the North American continent. Sometime during that day, all of the continental United States, much of Canada, and all of Mexico and the

Caribbean witness a partial solar eclipse. The eclipse begins at sunrise near Vancouver, British Columbia. The Moon's penumbral shadow sweeps slowly eastward, only to exit the coast of Newfoundland at sunset. Greatest eclipse, when more than two-thirds of the Sun is covered by the Moon, is witnessed from Baffin Island, in northern Canada.

Table 7.17 **Local Contact Times: 2000 December 25**
(Greatest eclipse: Time 17:35 U.T., Latitude 66° 20.2′N, Longitude 74° 6.3′W, Mag. 0.723)

City	P1	Maximum	P2	Magnitude
Atlanta	15:52	17:25	19:00	0.446
Boston	16:15	17:52	19:23	0.575
Chicago	15:44	17:17	18:53	0.548
Dallas	15:38	16:56	18:22	0.351
Detroit	15:51	17:26	19:03	0.570
Edmonton	15:30	16:44	18:03	0.526
Los Angeles	15:37	16:22	17:12	0.153
Miami	16:11	17:38	19:04	0.319
Montreal	16:09	17:45	19:18	0.614
New York City	16:09	17:47	19:20	0.559
Philadelphia	16:07	17:45	19:18	0.550
Toronto	15:58	17:34	19:09	0.593
Washington, D.C.	16:03	17:41	19:16	0.536

TOTAL SOLAR ECLIPSE OF 2001 JUNE 21 (SAROS 127)

The first solar eclipse of the twenty-first century spans the South Atlantic and the southern portion of Africa, and ends in the Indian Ocean (figure 7.5). Sunrise finds the Moon's umbral shadow first touching Earth far off the coast of Uruguay. Arcing northeastward, the shadow crosses the Atlantic, where greatest eclipse occurs several hundred kilometers out to sea.

The umbra first comes ashore in the village of Sumbe, in Angola. Next the umbra passes into Zambia, where the capital city of Lusaka witnesses 3 minutes 14 seconds of totality from its vantage point a little south of the center line. Crossing the Zambezi River into Zimbabwe, the shadow narrowly misses the capital of Harare, instead passing just to its north, straddling the border with Mozambique. The umbra leaves Mozambique at the seaside village of Chinde. Crossing the Mozambique Channel, the umbra

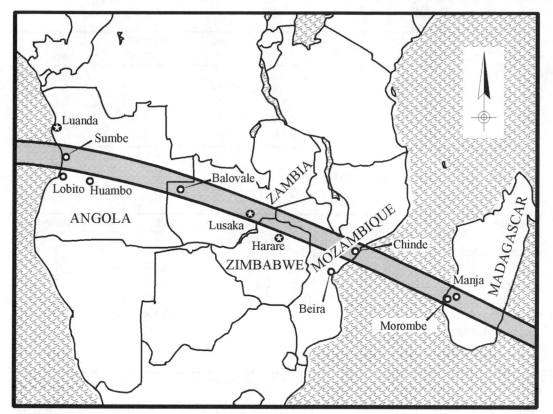

Figure 7.5 Track of the total solar eclipse of 2001 June 21.

flashes across Madagascar before passing into the Indian Ocean, where the eclipse ends at sunset.

Table 7.18 Total Solar Eclipse of 2001 June 21
Center Line Coordinate

U.T. hh:mm	Longitude °	′	Latitude °	′	Alt. °	Path Width km	Duration m:ss
10:37	−50	00	−36	37.1	00	127	2:06
10:39	−40	00	−31	44.2	10	137	2:29
10:46	−30	00	−26	20.9	21	150	2:59
11:00	−20	00	−20	42.3	34	165	3:39
11:23	−10	00	−15	28.2	47	182	4:22
11:54	00	00	−11	49.7	55	198	4:53
12:04	**+02**	**44.3**	**−11**	**15.8**	**55**	**200**	**4:57 (GE)**
12:27	+10	00	−10	49.3	52	198	4:48

Table 7.18 **(Continued)**

U.T. hh:mm	Longitude °	′	Latitude °	′	Alt. °	Path Width km	Duration m:ss
12:54	+20	00	−12	24.5	42	183	4:11
13:13	+30	00	−15	43.9	29	165	3:28
13:24	+40	00	−19	55.3	17	150	2:51
13:29	+50	00	−24	24.6	05	136	2:22

Table 7.19 **Local Contact Times: 2001 June 21**

   a. Total

City/town	P1	U1	Maximum	U2	P2	Duration m:ss
Sumbe, Angola	10:57	12:36	12:38	12:40	14:08	4:35
Balovale, Zambia	11:26	12:59	13:01	13:03	14:22	3:46
Lusaka, Zambia	11:41	13:09	13:10	13:12	14:26	3:15
Chinde, Mozambique	12:00	13:20	13:21	13:22	14:31	2:52
Morombe, Madagascar	12:12	13:25	13:26	13:28	14:32	2:23
Manja, Madagascar	12:13	13:26	13:27	13:29	14:32	2:04

   b. Partial

City	P1	Maximum	P2	Magnitude
Cape Town	11:17	12:37	13:49	0.516
Johannesburg	11:39	13:03	14:17	0.737
Nairobi, Kenya	12:15	13:25	14:26	0.523
Port Elizabeth, South Africa	11:35	12:52	14:01	0.552

Table 7.20 **Weather Prospects: 2001 June 21**

City	Clear	Scattered	Broken	Overcast
Beira, Mozambique	10.1	65.2	20.5	4.1
Harare, Zimbabwe	23.0	46.0	30.6	0.4
Lobito, Angola	Clouds < 3/10 = 24.5 days		Precip > 0.1 inch = 0 days	
Lusaka, Zambia	18.8	43.1	36.7	1.4

ANNULAR SOLAR ECLIPSE OF 2001 DECEMBER 14 (SAROS 132)

For those who are thinking about the cold that the coming winter promises, this annular eclipse offers the perfect attitude adjustment. The track of annularity first strikes Earth far from land, in the middle of the Pacific Ocean. Heading eastward, the central-eclipse path passes about 480 kilometers (300 miles) south of the Hawaiian islands, where a 0.842-magnitude eclipse is seen in Honolulu that morning. The track continues toward the southeast over open water, where greatest eclipse occurs far from land, but nearly on the Equator.

Veering northeast, the path of annularity finally strikes land in Central America only as the Sun nears the horizon. As shown in figure 7.6, several towns and cities in Costa Rica see annularity just before sunset in the late afternoon. The capital city of San José has an especially interesting view of Baily's Beads, as it rides on the southern limit of annularity.

Except for portions of New England, upstate New York, and Alaska, the United States sees a partial solar eclipse on December 14. East of the Mississippi River, sunset comes before the eclipse is finished, while the western half of the country has an uninterrupted view.

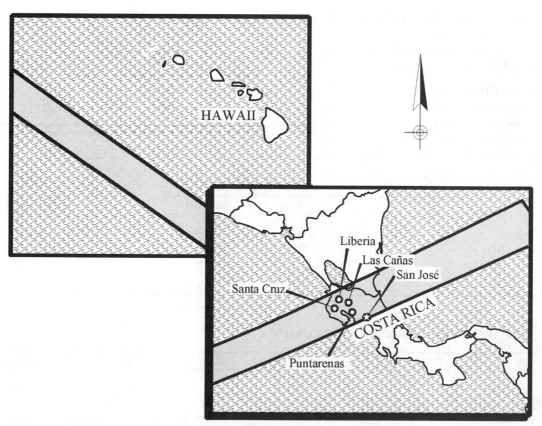

Figure 7.6   Path of the annular solar eclipse of 2001 December 14.

Table 7.21   **Annular Solar Eclipse of 2001 December 14**
  Center Line Coordinates

U.T. hh:mm	Longitude °	 ′	Latitude °	 ′	Alt. °	Path Width km	Duration m:ss
19:10	180	00	+28	05.7	04	171	3:08
19:14	−170	00	+22	56.2	15	154	3:14
19:24	−160	00	+17	01.0	29	142	3:19
19:45	−150	00	+10	34.1	44	130	3:30
20:16	−140	00	+04	29.9	59	124	3:43
20:52	**−130**	**41.4**	**+00**	**37.3**	**+66**	126	3:53 (GE)
20:54	−130	00	+00	26.2	66	126	3:53
21:31	−120	00	−00	35.8	57	131	3:51
21:59	−110	00	+00	57.4	43	138	3:41
22:18	−100	00	+04	09.1	29	148	3:30
22:29	−90	00	+08	10.2	16	161	3:21
22:33	−80	00	+12	29.7	04	172	3:16

Table 7.22   **Local Contact Times: 2001 December 14**
  a. Annular

City/town	P1	U1	Maximum	U2	P2	Duration m:ss
Santa Cruz, Costa Rica	21:11	22:30	22:31	22:33	23:39	3:11
Puntarenas, Costa Rica	21:12	22:31	22:32	22:33	23:39	2:55
Liberia, Costa Rica	21:11	22:30	22:31	22:33	23:39	2:35
Las Canas, Costa Rica	21:12	22:30	22:32	22:33	23:39	3:15
San José, Costa Rica	21:13	22:32	22:32	22:33	23:39	1:17

  b. Partial

City	P1	Maximum	P2	Magnitude
Dallas	20:55	22:02	23:00	0.319
Honolulu	18:08	19:26	21:00	0.842
Los Angeles	20:04	21:10	22:12	0.199
Mexico City	20:44	22:12	23:26	0.590

Table 7.23   **Weather Prospects: 2001 December 14**

City	Clear	Scattered	Broken	Overcast
Honolulu	11.8	41.0	28.6	18.7
Puntarenas, Costa Rica	28.3	26.5	32.9	12.3

ANNULAR SOLAR ECLIPSE OF 2002 JUNE 10–11 (SAROS 137)

Sometimes the greatest challenge to seeing a solar eclipse is getting to the path of totality or annularity. This eclipse is a case in point. Beginning in the Celebes Sea, just north of the Indonesian island of Sulawesi, the track cuts a long, wet track across the Pacific (figure 7.7). Apart from passing over the small South Pacific islands of Sangihe and Talaud, the path of annularity never touches land! Instead, it spans nearly the full breadth of the Pacific, ending just off the coast of Cabo San Lucas and Puerto Vallarta, Mexico.

Although the view of annularity may be witnessed by relatively few people, a wonderful photo opportunity (like that captured in figure 7.8) exists for those who visit either Puerto Vallarta or Cabo San Lucas, Mexico. There, the eclipse will reach magnitude 0.974 just as the Sun touches the

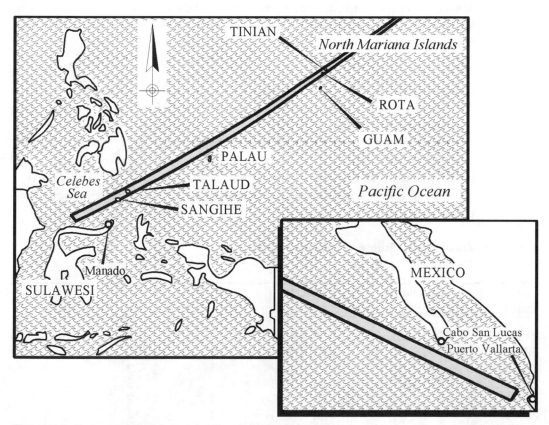

Figure 7.7 Path of the annular solar eclipse of 2002 June 10.

Figure 7.8 Frans Pyck captured the exquisite beauty of the 1992 January 4 sunset annular eclipse, a scene that many will see again during the June 2002 event. (85-mm lens at f/8, 1/500th-second exposures, and ISO 100 film. No solar filter was needed because of the deep filtering effect of the clouds.)

western horizon. For many, the return to Cabo will bring back memories of 1991 July 11, when the southern tip of the Baja California peninsula saw a spectacular total eclipse.

While annularity will elude many of us, a large portion of the Pacific rim will witness a partial eclipse. Sometime during the day, the Moon will cover a portion of Sun for observers in the western half of North America, all of Japan, much of China and the Far East, and even northeastern Australia.

Table 7.24 **Annular Solar Eclipse of 2002 June 10–11**
Center Line Coordinates

U.T. hh:mm	Longitude °	′	Latitude °	′	Alt. °	Path Width km	Duration m:ss
21:56	+130	00	+05	41.3	11	67	1:07
22:04	+140	00	+11	16.4	24	53	0:58
22:19	+150	00	+17	40.9	38	39	0:48
22:42	+160	00	+24	19.8	54	26	0:37
23:11	+170	00	+30	06.5	69	17	0:28
23:40	180	00	+34	08.0	78	14	0:23
23:45	**−178**	**36.7**	**+34**	**32.7**	**78**	**13**	**0:23 (GE)**

Table 7.24 **(Continued)**

U.T. hh:mm	Longitude °	Longitude ′	Latitude °	Latitude ′	Alt. °	Path Width km	Duration m:ss
00:07	−170	00	+36	15.1	73	14	0:24
00:30	−160	00	+36	40.5	62	18	0:28
00:51	−150	00	+35	39.0	51	24	0:35
01:08	−140	00	+33	24.1	39	32	0:43
01:20	−130	00	+30	10.8	28	43	0:50
01:29	−120	00	+26	17.3	16	53	0:57
01:33	−110	00	+22	02.6	05	53	0:57

Table 7.25 **Local Contact Times: 2002 June 10–11**

a. Annular

City/town	P1	U1	Maximum	U2	P2	Duration m:ss
Tahuna, Sangihe Island	—	21:54	21:55	21:55	22:59	1:04

b. Partial

City	P1	Maximum	P2	Magnitude
Anchorage	23:34	00:27	01:19	0.284
Cabo San Lucas, Mexico	00:28	01:33	—	0.974
Denver	00:20	01:16	02:07	0.508
Guam	21:01	22:09	23:29	0.975
Juneau	23:49	00:42	01:33	0.305
Las Vegas	00:14	01:20	02:18	0.690
Los Angeles	00:13	01:21	02:23	0.773
Portland, Oregon	00:02	01:06	02:04	0.534
San Francisco	00:06	01:16	02:18	0.721
Seattle	00:02	01:03	02:00	0.482
Tokyo	21:41	22:40	23:45	0.454
Tucson	00:20	01:24	02:22	0.744

Table 7.26 **Weather Prospects: 2002 June 10–11**

City	Clear	Scattered	Broken	Overcast
Palau	0	12.2	13.9	73.9
Manado, Sulawesi	0.9	47.6	42.5	9.0

TOTAL SOLAR ECLIPSE OF 2002 DECEMBER 4 (SAROS 142)

A single location on Earth's surface may expect to see a total solar eclipse once in three centuries on average, but the west coast of the African country of Angola is well ahead of those odds. Just eighteen months before this event, a total eclipse struck land over the town of Sumbe on its way toward Zambia, Zimbabwe, and Mozambique. Now again, Sumbe braces for another encounter with the shadow of the Moon, though it will not be a direct hit this time. Instead, the Moon's shadow barely misses the town, skimming just a handful of kilometers to the south (figure 7.9). Though the track may be similar to the June 2001 eclipse, both events are decidedly different. The 2001 eclipse offered nearly five minutes of totality, but here it will last no more than about two minutes.

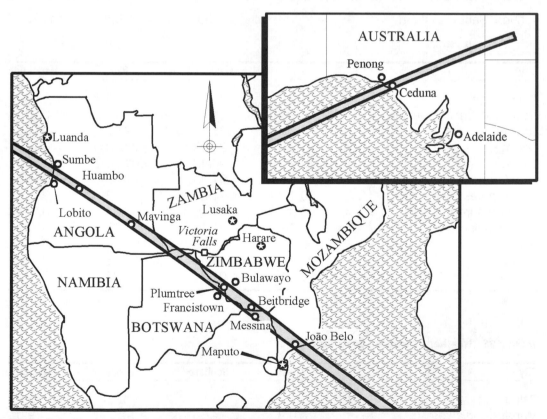

Figure 7.9 Track of the total solar eclipse of 2002 December 4.

Passing inland, the shadow passes over several small towns in Angola, including the city of Huambo. Residents there witness 53 seconds of totality not long after sunrise that day. From Angola, just as it did a year and a half earlier, the shadow enters Zambia, though farther to the south this time. Barely missing Victoria Falls (wouldn't that have been a photo opportunity!), the eclipse track slices briefly into the panhandle of Namibia before skirting along the Botswana-Zimbabwe border. The shadow finally crosses southern Mozambique before leaving the African continent for the Indian Ocean.

The path's landfall is not completely over yet, however, as the track makes its way toward Australia shortly before leaving the Earth's surface entirely. The coastal town of Ceduna, in South Australia, stands to witness an especially spectacular sunset eclipse as the narrow track prepares to leave the Earth's surface.

Table 7.27 **Total Solar Eclipse of 2002 December 4**
Center Line Coordinates

U.T. hh:mm	Longitude °	′	Latitude °	′	Alt. °	Path Width km	Duration m:ss
05:50	00	00	−04	39.9	02	34	0:26
05:53	+10	00	−09	27.2	14	47	0:43
06:02	+20	00	−15	19.2	27	62	1:01
06:18	+30	00	−22	08.4	41	75	1:23
06:41	+40	00	−29	11.3	56	83	1:44
07:07	+50	00	−35	15.7	68	87	1:59
07:32	**+60**	**07.0**	**−39**	**13.4**	**72**	**87**	**2:04 (GE)**
07:54	+70	00	−42	15.4	68	84	2:00
08:12	+80	00	−43	25.9	60	80	1:50
08:29	+90	00	−43	22.4	51	75	1:37
08:43	+100	00	−42	13.8	42	68	1:21
08:54	+110	00	−40	07.3	32	61	1:07
09:03	+120	00	−37	10.6	22	51	0:51
09:09	+130	00	−33	34.4	12	40	0:36
09:11	+140	00	−29	32.3	02	29	0:24

Table 7.28 **Local Contact Times: 2002 December 4**
a. Total

City/town	P1	U1	Maximum	U2	P2	Duration m:ss
Huambo, Angola	04:59	05:57	05:57	05:58	07:03	0:53
Mavinga, Angola	05:02	06:02	06:03	06:03	07:12	0:34

Table 7.28 **(Continued)**

City/town	P1	U1	Maximum	U2	P2	Duration m:ss
Plumtree, Zimbabwe	05:08	06:13	06:14	06:14	07:28	1:15
Beitbridge, Zimbabwe	05:11	06:18	06:18	06:19	07:35	1:21
Messina, South Africa	05:11	06:18	06:18	06:19	07:35	1:08
Joao Belo, Mozambique	05:17	06:26	06:26	06:27	07:45	0:59
Ceduna, Australia	08:10	09:10	09:10	09:10	10:05	0:32

b. Partial

City	P1	Maximum	P2	Magnitude
Cape Town	05:31	06:29	07:32	0.591
Johannesburg	05:17	06:23	07:37	0.886
Perth, Australia	07:58	09:06	10:07	0.823

Table 7.29 **Weather Prospects: 2002 December 4**

City	Clear	Scattered	Broken	Overcast
Adelaide, Australia	8.2	57.6	29.4	4.7
Bulawayo, Zimbabwe	1.9	30.0	58.6	9.5
Francistown, Botswana	5.2	35.6	30.4	28.8
Lobito, Angola	Clouds <3/10 = not recorded		Precip >0.1 inch = 0.4 days	
Maputo, Mozambique	4.5	31.1	44.6	19.8

ANNULAR SOLAR ECLIPSE OF 2003 MAY 31 (SAROS 147)

This is an eclipse for the twenty-first-century record books! As you can tell by the accompanying map (figure 7.10), the path of this annular eclipse is unusually broad. Because of the oblique angle of the Moon's shadow to the Earth, the path spans some 4,900 kilometers (3,000 miles), broader than any other eclipse path in this century! Sad that its remote location and low altitude will allow relatively few eclipse chasers to witness the event.

Indeed, while annularity will be visible from northernmost Scotland, the Shetland Islands, Iceland, and Greenland, the Sun will be barely above the horizon when maximum eclipse strikes. Still, if you can find a near-perfect horizon, you will see a spectacular ring of fire.

The cities of Inverness, Scotland, and Reykjavik, Iceland, will both see the Sun rise as a thin ring of light that morning, with dawn and annularity coinciding almost perfectly. But, again, the Sun's altitude is working

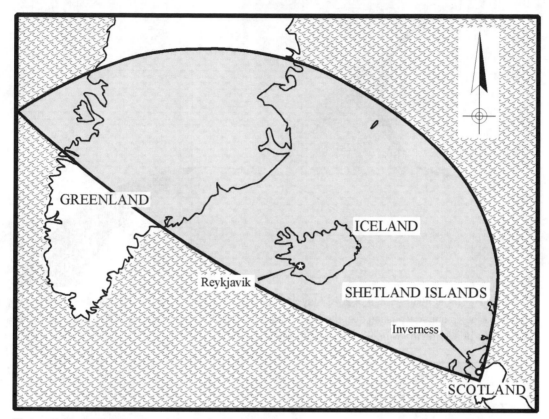

Figure 7.10 Path of the annular solar eclipse of 2003 May 31.

against us. From Inverness it will be right on the horizon during the 1 minute 15 seconds of annularity, while from Reykjavik it will stand a mere 3° above the horizon at maximum eclipse, which will last for 3 minutes 36 seconds. This nearly matches the point of greatest eclipse, which is set in the Denmark Strait, between Iceland and Greenland.

The broad nature of annularity's track means that a wide range of northern locales, primarily in Russia, Alaska, and the Yukon territory of Canada, will witness a partial eclipse sometime during the day. Canadian and Alaskan observers should note, however, that for them the eclipse will be seen on the afternoon of May 30.

Take a look at the coordinates in table 7.30. Notice anything odd or unusual about them? The shadow will appear to be moving backwards (i.e., east to west, rather than west to east). No, the Moon is not going to reverse suddenly in its orbit. Rather, this strange effect is the result of the shadow crossing the Earth's surface *beyond* the North Pole (i.e., after crossing over the top of the pole; see figure 7.11).

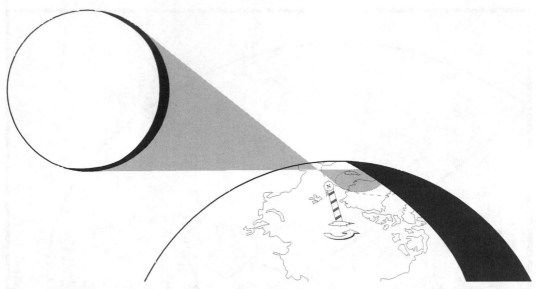

Figure 7.11 The annular eclipse of 2003 May 31 is perhaps the most unusual event in this chapter. Because the Moon's shadow passes over the North Pole before it strikes the Earth's surface, the path of the eclipse will actually run in reverse—from east to west!

Table 7.30 **Annular Solar Eclipse of 2003 May 31**
Center Line Coordinates

U.T. hh:mm	Longitude °	′	Latitude °	′	Alt. °	Path Width km	Duration m:ss
04:01	−20	00	+62	59.0	01	4900	3:36
04:03	−20	00	+64	45.0	02	4600	3:37
04:08	−24	00	+66	40.0	03	4200	3:37
04:09	**−24**	**00.2**	**+66**	**49.4**	**03**	**4498**	**3:37 (GE)**
04:10	−28	00	+67	10.0	03	4200	3:37
04:13	−32	00	+66	59.6	02	4500	3:36

Table 7.31 **Local Contact Times: 2003 May 31**
a. Annular

City/town	P1	U1	Maximum	U2	P2	Duration m:ss
Reykjavik, Iceland	—	04:02	04:04	04:05	05:01	3:36
Lerwick, Shetland Islands	—	03:46	03:47	03:47	04:45	1:27
Inverness, Scotland	—	03:44	03:45	03:46	04:42	1:15

Table 7.31 **(Continued)**

  b. Partial

City	P1	Maximum	P2	Magnitude
Fairbanks	04:23	05:21	06:17	0.587
Moscow	04:24	05:24	06:29	0.774
St. Petersburg	04:33	05:34	06:38	0.833
Whitehorse, Yukon, Canada	04:26	05:22	—	0.610

If you compare the above values with those in table 7.30, you will find a discrepancy in the times. Table 7.30 shows that the first contact point will occur at a little past 04:00 U.T., but in table 7.31, annularity strikes Scotland and the Shetland Islands some fifteen minutes earlier. Once again, the difference is caused by the eclipse's unusual geometry. Though the umbra's central line contacts the Earth's surface at 04:01 U.T., the shadow's edge strikes our planet some fifteen minutes earlier (and exits about fifteen minutes after).

Table 7.32 **Weather Prospects: 2003 May 31**

City	Clear	Scattered	Broken	Overcast
Inverness, Scotland	1.6	24.1	52.4	21.4
Lerwick, Shetland Islands	0.6	21.2	39.1	32.7
Reykjavik, Iceland	0.9	26.9	35.2	36.3

TOTAL SOLAR ECLIPSE OF 2003 NOVEMBER 23 (SAROS 152)

If you thought the first solar eclipse of 2003 was tough to get to, wait until you see where this one is visible from! Though broad, the path of totality only crosses one continent: Antarctica! Beginning in the region known as Enderby Land, adjacent to the Atlantic Ocean, the track hooks eastward across the American Highlands before passing into the Indian Ocean well to the south of Australia (figure 7.12). The only inhabited spot found along the entire track is the Russian scientific research station Mirnyy. Unfortunately, totality there, or anywhere along the entire path, never reaches two minutes in length. If you travel to see this eclipse, you are a *real* die-hard!

While totality will be all but inaccessible, most of Australia and New Zealand's South Island will be treated to a partial eclipse that morning.

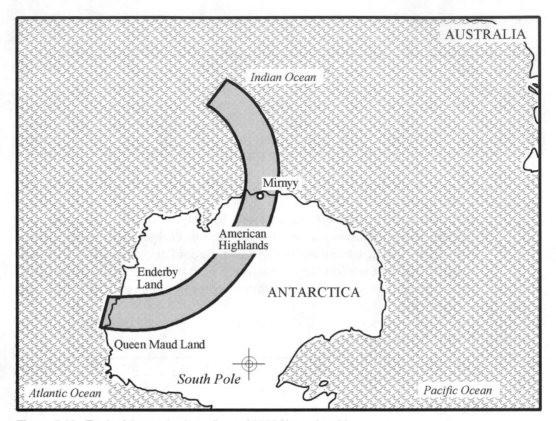

Figure 7.12 Track of the total solar eclipse of 2003 November 23.

Table 7.33 **Total Solar Eclipse of 2003 November 23**
Center Line Coordinates

U.T. hh:mm	Longitude °	′	Latitude °	′	Alt. °	Path Width km	Duration m:ss
22:25	+90	00	−57	33.9	07	543	1:44
22:47	+90	00	−71	34.9	15	497	1:57
22:50	**+88**	**18.3**	**−72**	**39.3**	**15**	**496**	**1:57 (GE)**
22:55	+80	00	−75	38.8	14	485	1:56
23:00	+70	00	−77	20.4	13	482	1:55
23:03	+60	00	−78	06.4	12	483	1:53
23:06	+50	00	−78	16.0	11	484	1:52
23:08	+40	00	−77	51.9	10	488	1:49
23:11	+30	00	−76	43.3	08	494	1:45
23:14	+20	00	−73	58.5	05	502	1:41

Table 7.34 **Local Contact Times: 2003 November 23**
      a. Total

City/town	P1	U1	Maximum	U2	P2	Duration m:ss
Mirnyy Station, Antarctica	21:47	22:37	22:38	22:39	23:31	1:54

b. Partial

City	P1	Maximum	P2	Magnitude
Adelaide, Australia	20:55	21:40	22:29	0.429
Canberra	21:01	21:44	22:30	0.312
Melbourne	21:00	21:47	22:37	0.406
Perth, Australia	20:54	21:39	22:27	0.608
Sydney	21:02	21:42	22:25	0.257
Wellington, New Zealand	21:52	22:16	22:41	0.056

PARTIAL SOLAR ECLIPSE OF 2004 APRIL 19 (SAROS 119)

With only two partial eclipses on the roster, 2004 will prove rather disappointing for solar-eclipse fans. While more than two-thirds of the Sun will be eclipsed at maximum during this first event, that point lies inaccessibly close to the Antarctic coast, thousands of kilometers south of the Cape of Good Hope. Viewers in the southern third of Africa, south of a diagonal line stretching from approximately northern Angola to Mozambique, will also see the partial eclipse, although it will be comparatively shallow. The deepest eclipse will be visible from points immediately adjacent to Cape Town, South Africa, where the Moon will cover approximately half of the solar disk that afternoon.

Table 7.35 **Local Contact Times: 2004 April 19**
(Greatest eclipse: Time 13:35 u.t., Latitude −61° 35.6′, Longitude −44° 20.6′, Mag. 0.735)

City	P1	Maximum	P2	Magnitude
Cape Town	12:52	14:11	15:22	0.508
Johannesburg	13:25	14:34	15:35	0.417
Pretoria	13:27	14:35	15:36	0.410
Windhoek, Namibia	13:23	14:31	15:32	0.330

PARTIAL SOLAR ECLIPSE OF 2004 OCTOBER 14 (SAROS 124)

Nearly six months and 180° in longitude later, the second partial eclipse of 2004 will blanket portions of northeastern Asia, including all of Japan, northeastern Mongolia and China, and most of Siberia. Observers in the western half of Alaska will witness the greatest eclipse just as the Sun sets, because of the International Date Line, on the afternoon of October 13. The point of greatest eclipse lies near the town of Kenai, southwest of Anchorage. There, more than 90 percent of the Sun will be eclipsed at sunset. Other Alaskan towns, including Nome, Kotzebue, and Bethel will also see the Sun drop below the southwestern horizon still in eclipse.

Table 7.36 **Local Contact Times: 2004 October 14**
(Greatest eclipse: Time 03:00 u.t., Latitude 61° 15.0′, Longitude −153° 36.6′, Mag. 0.928)

City	P1	Maximum	P2	Magnitude
Bethel, Alaska	01:50	02:57	—	0.924
Kenai, Alaska	01:56	—	—	0.928
Kotzebue, Alaska	01:39	02:45	—	0.912
Kyoto	01:49	02:35	03:22	0.157
Osaka	01:51	02:35	03:20	0.147
Nome, Alaska	01:41	02:49	—	0.914
Tokyo	01:45	02:40	03:36	0.240

ANNULAR/TOTAL SOLAR ECLIPSE OF 2005 APRIL 8 (SAROS 129)

This eclipse is an example of the rarest type of solar eclipse: an annular-total eclipse. Normally the Moon appears either slightly larger than the Sun (creating a total solar eclipse) or slightly smaller (producing an annular eclipse). But at those rare times when the Sun and Moon appear very nearly the *exact* same size, then an annular eclipse will occur at the extreme ends of the eclipse path, while a total eclipse will occur around the center of the track. Figure 3.13 already showed that it is possible to capture hints of the inner corona, chromosphere, and prominences on film even as the photosphere peeks around lunar peaks and valleys.

The nature of annular-total eclipses means that an observer must be located precisely within the path; there is very little margin for error. As mentioned in this chapter's opening paragraphs, use the numbers here as a general guide only; as the eclipse nears, rely on the latest data to determine the exact track.

The eclipse first reaches the Earth in the Pacific Ocean, just east-southeast of New Zealand. Greatest eclipse occurs well out in the South Pacific. Traveling across a vast expanse of ocean, greatest eclipse, featuring 42 seconds of totality, will be reached far out at sea. Having changed back to an annular eclipse, the path strikes land for the first time in Central America, at the border of Costa Rica and Panama (figure 7.13). The thread of annularity misses the Panamanian city of David, but passes over the tiny coastal town of Pedregal, less than 16 kilometers (10 miles) to the south. It also dashes over or near Las Lajas, Penonome, Anton, and San Carlos, all found on the road from David to Panama City.

The path then hops across the Gulf of Urabá and into northern Colombia, where it crosses over Pueblonuevo, about 40 kilometers (25 miles) to the southeast of the city of Montería. As sunset draws near, the path moves

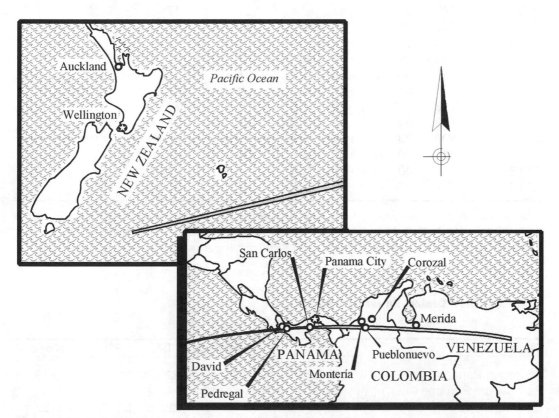

Figure 7.13   Path of the annular/total solar eclipse of 2005 April 8.

into Venezuela, passing south of the city of Mérida shortly before leaving the Earth's surface.

Narrow as the maximum-eclipse track will be, a partial eclipse will be seen across a broad portion of Earth's surface. New Zealand's North Island is just tucked inside the partial-eclipse region, and is set to witness the Sun rising already at maximum eclipse. On the other side of the Pacific, an afternoon eclipse will be seen from much of the west coast of South America, all of Central America, and along the southern and mid-Atlantic regions of the United States.

Table 7.37 **Annular/Total Solar Eclipse of 2005 April 8**
Center Line Coordinates

U.T. hh:mm	Longitude °	′	Latitude °	′	Alt. °	Path Width[1] km	Duration m:ss	Type[2]
18:53	180	00	−47	16.3	03	23	0:24	A
18:55	−170	00	−45	12.8	11	17	0:18	A
19:00	−160	00	−42	12.0	19	8	0:10	A
19:08	−150	00	−37	54.1	29	1	0:02	A
19:23	−140	00	−31	46.4	42	11	0:16	T
19:50	−130	00	−23	01.5	57	21	0:32	T
20:30	−12	00	−11	45.7	69	27	0:42	T
20:36	**−118**	**57.7**	**−10**	**34.6**	**70**	**27**	**0:42 (GE)**	**T**
21:13	−110	00	−01	44.8	61	23	0:34	T
21:43	−100	00	+04	20.1	45	11	0:15	T
22:01	−90	00	+07	23.7	31	2	0:02	A
22:12	−80	00	+08	30.4	18	14	0:16	A
22:17	−70	00	+08	17.9	07	25	0:27	A

1. Take a close look at how the eclipse path seesaws in width, first narrowing down, then expanding, only to grow narrow again. These changes reflect the eclipse oscillating from an annular eclipse to total, then back to annular again.

2. This column shows the type of eclipse (*A* for annular, *T* for total) occurring at each set of coordinates.

Table 7.38 **Local Contact Times: 2005 April 8**
a. Annular/Total

City/town	P1	U1	Maximum	U2	P2	Duration m:ss
Pedregal, Panama	20:52	22:10	22:10	22:10	23:17	0:13
San Carlos, Panama	20:56	22:12	22:12	22:12	23:18	0:15
Pueblonuevo, Colombia	21:02	22:14	22:15	22:15	23:18	0:21

Table 7.38 **(Continued)**

b. Partial

City	P1	Maximum	P2	Magnitude
Atlanta	21:35	22:18	22:58	0.209
Auckland, New Zealand	—	05:49	06:49	0.656
Houston	21:17	22:11	23:01	0.295
Miami	21:20	22:19	23:14	0.466
New Orleans	21:21	22:15	23:05	0.314
Panama City, Panama	20:57	22:12	23:18	0.985

Table 7.39 **Weather Prospects: 2005 April 8**

City	Clear	Scattered	Broken	Overcast
Corozal, Colombia	0	30.7	27.2	42.1
David, Panama	0	21.8	32.7	45.5
Montería, Colombia	1.7	29.9	59.0	9.4

ANNULAR SOLAR ECLIPSE OF 2005 OCTOBER 3 (SAROS 134)

The Moon's shadow returns to Earth for a visit on 2005 October 3, this time to the Iberian peninsula and Africa (figure 7.14). The umbra never quite makes it to the ground, however; instead, witnesses will see a broad annular eclipse as the too-small Moon passes in front of the Sun.

After touching down in the Atlantic, the shadow first makes landfall near the border shared by Portugal and Spain. The cities of Vigo, Spain, and Braga, Portugal, are among the first to see annularity. The citizens of Madrid see a ringlike Sun for 4 minutes 11 seconds, while Valencia, closer to the edge of annularity, enjoys a 2-minute-6-second central eclipse.

Crossing the Mediterranean, the path strikes Algeria, where the capital city of Algiers witnesses a 3-minute-50-second annular eclipse. Unfortunately, few sizable towns lie farther along the eclipse track as the Moon's shadow journeys through Tunisia, Libya, northeastern Chad, Sudan, southwestern Ethiopia, Kenya, and finally Somalia. Greatest eclipse, with 4 minutes 31 seconds of annularity, occurs in Sudan, northeast of Salim. Later, Marsabit, Kenya, will witness more than four minutes of annularity, while

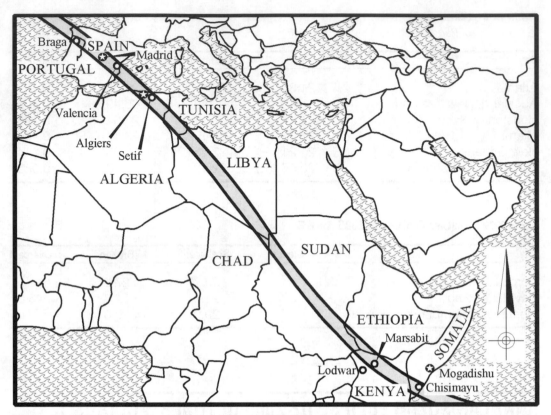

Figure 7.14 Path of the annular solar eclipse of 2005 October 3.

Chisimayu, on the Somalian coast, will enjoy an annular phase lasting 2 minutes 51 seconds before the Moon's shadow dives into the Indian Ocean.

Table 7.40 **Annular Solar Eclipse of 2005 October 3**
Center Line Coordinates

U.T. hh:mm	Longitude °	′	Latitude °	′	Alt. °	Path Width km	Duration m:ss
08:43	−30	00	+47	17.8	06	215	3:55
08:46	−20	00	+45	33.5	14	204	4:03
08:52	−10	00	+42	50.9	22	193	4:08
09:02	00	00	+38	48.3	32	183	4:14
09:19	+10	00	+32	46.4	45	172	4:22
09:50	+20	00	+23	39.0	61	162	4:28
10:32	**+28**	**45.0**	**+12**	**52.4**	**71**	**162**	**4:31 (GE)**

Table 7.40   **(Continued)**

U.T. hh:mm	Longitude °	′	Latitude °	′	Alt. °	Path Width km	Duration m:ss
10:38	+30	00	+11	17.5	70	163	4:32
11:23	+40	00	+00	43.2	56	178	4:27
11:52	+50	00	−05	22.8	39	191	4:17
12:08	+60	00	−08	28.9	25	200	4:07
12:17	+70	00	−09	44.3	13	207	3:58
12:20	+80	00	−09	45.0	03	217	3:48

Table 7.41   **Local Contact Times: 2005 October 3**

   a. Annular

City/town	P1	U1	Maximum	U2	P2	Duration m:ss
Braga, Portugal	07:38	08:52	08:53	08:54	10:16	2:36
Madrid	07:40	08:55	08:57	08:59	10:23	4:11
Valencia, Spain	07:42	09:01	09:02	09:03	10:30	2:06
Algiers	07:44	09:04	09:06	09:08	10:37	3:50
Setif, Algeria	07:46	09:08	09:10	09:12	10:41	4:04
Marsabit, Kenya	09:31	11:13	11:16	11:18	12:51	4:27
Chisimayu, Somalia	09:48	11:29	11:30	11:32	13:01	2:51

   b. Partial

City	P1	Maximum	P2	Magnitude
Athens	08:09	09:34	11:03	0.637
Johannesburg	11:01	11:50	12:37	0.146
London	06:48	08:01	09:18	0.663
Munich	07:56	09:11	10:32	0.613
Paris	07:47	09:02	10:22	0.702
Rome	07:53	09:15	10:43	0.734

Table 7.42   **Weather Prospects: 2005 October 5**

City	Clear	Scattered	Broken	Overcast
Algiers	10.6	40.6	42.0	6.8
Lodwar, Kenya	0.5	36.2	61.6	1.7
Madrid	28.4	34.7	25.0	11.8
Mogadishu, Somalia	0.6	52.3	43.2	4.0
Valencia, Spain	21.4	42.6	25.7	10.2

TOTAL SOLAR ECLIPSE OF 2006 MARCH 29 (SAROS 139)

The Moon's shadow returns to Africa less than six months later, this time bringing with it a total solar eclipse (figure 7.15). After touching down at sunrise in easternmost Brazil, where the coastal city of Natal and adjacent communities see the Sun rise above the Atlantic in total eclipse, the umbra races across the ocean for a midmorning appointment with Ghana's capital city of Accra. Traveling inland, the umbra slices through Togo and Benin, and into northwestern Nigeria. The towns of Gusau and Katsina, found near the umbra's center line, behold more than 3 minutes 50 seconds of totality, given clear skies.

Father along, the shadow traverses the barren region of eastern Niger, then quickly passes into northern Chad and eastern Libya. Greatest eclipse, where totality lasts for 4 minutes 7 seconds, occurs along their common (albeit inaccessible) border. Notice how the track of this eclipse crosses

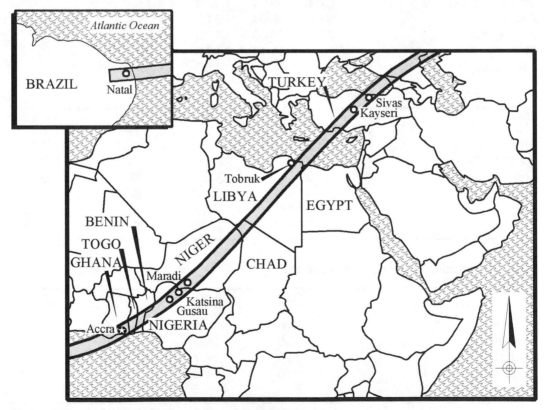

Figure 7.15   Track of the total solar eclipse of 2006 March 29.

paths with that of 2005 October 3 in the barren Libyan desert some 200 kilometers (130 miles) east of the town of Waw al-Kabir. Ironically, though the uninhabited desert handily beats the "once-every-300-years" rule, the town itself is outside of the maximum-eclipse path both times!

Turkey will be next on the shadow's schedule. Passing east of Ankara, totality will strike the smaller cities of Kayseri and Sivas, where totality will last for more than two minutes. After a short hop over the Black Sea, totality next strikes the northwest corner of Georgia, and then moves across Kazakhstan before leaving Earth as the Sun sets.

As the umbral shadow slides across central Africa, a partial eclipse is witnessed across all but the southernmost corner of the continent as well as all of Europe, the Middle East, and western Asia.

Table 7.43 **Total Solar Eclipse of 2006 March 29**
Center Line Coordinates

U.T. hh:mm	Longitude °	′	Latitude °	′	Alt. °	Path Width km	Duration m:ss
08:36	−30	00	−05	36.5	07	139	2:08
08:41	−20	00	−03	34.8	19	155	2:30
08:52	−10	00	+00	06.8	32	172	2:59
09:12	00	00	+06	14.3	47	185	3:30
09:45	+10	00	+15	34.3	63	188	4:00
10:12	**+16**	**45.9**	**+23**	**08.6**	**67**	**194**	**4:07 (GE)**
10:23	+20	00	+26	39.3	66	180	4:05
10:53	+30	00	+35	46.3	56	171	3:48
11:13	+40	00	+42	03.2	44	163	3:25
11:26	+50	00	+46	15.2	35	156	3:04
11:35	+60	00	+49	01.2	26	149	2:44
11:40	+70	00	+50	44.5	19	143	2:30
11:44	+80	00	+51	38.6	12	137	2:14
11:46	+90	00	+51	51.8	06	131	2:01

Table 7.44 **Local Contact Times: 2006 March 29**
a. Total

City/town	P1	U1	Maximum	U2	P2	Duration m:ss
Natal, Brazil	—	08:34	08:35	08:36	09:34	1:30
Accra, Ghana	08:00	09:09	09:11	09:12	10:29	3:04
Gusau, Nigeria	08:17	09:31	09:33	09:35	10:54	3:51
Katsina, Nigeria	08:20	09:34	09:36	09:37	10:58	3:54

Table 7.44 **(Continued)**

City/town	P1	U1	Maximum	U2	P2	Duration m:ss
Maradi, Niger	08:20	09:35	09:36	09:37	10:58	2:47
Kayseri, Turkey	09:47	11:03	11:04	11:05	12:19	1:55
Sivas, Turkey	09:50	11:06	11:07	11:09	12:21	2:23

b. Partial

City	P1	Maximum	P2	Magnitude
Ankara, Turkey	09:45	11:02	12:17	0.973
Cairo	09:27	10:47	12:06	0.864
London	09:45	10:33	11:21	0.275
Madrid	09:17	10:12	11:09	0.357
Moscow	10:10	11:15	12:18	0.651
Paris	09:39	10:32	11:26	0.338
Rome	09:27	10:36	11:45	0.595

Table 7.45 **Weather Prospects: 2006 March 29**

City	Clear	Scattered	Broken	Overcast
Accra, Ghana	3.6	15.8	78.4	2.2
Maradi, Niger	0	13.2	81.4	5.1
Natal, Brazil	1.5	37.5	45.7	15.3
Sivas, Turkey	18.4	20.3	40.7	20.6
Tobruk, Libya	17.1	40.4	29.5	13.0

ANNULAR SOLAR ECLIPSE OF 2006 SEPTEMBER 22 (SAROS 144)

The broad path of this annular eclipse slides from northeastern South America southward along the Atlantic toward the Cape of Good Hope, though only those eclipse chasers in the air or on the sea stand much chance of seeing the spectacle.

A stunning annular sunrise is seen from northeastern Guyana as well as northern Suriname and French Guiana (figure 7.16). The capital cities of Paramaribo, Suriname, and Cayenne, French Guiana, each see annularity begin within minutes of dawn, as does Kourou, the European Space Agency's launch site in French Guiana. The view from Cayenne, with its

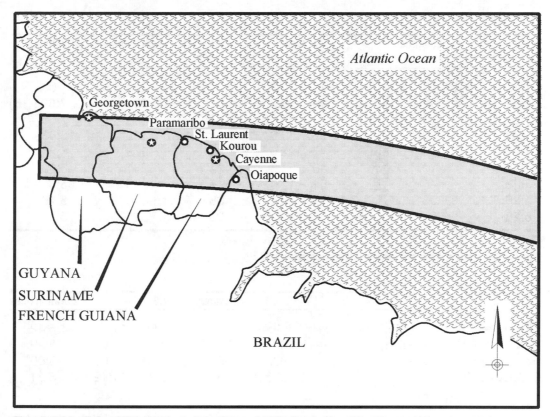

Figure 7.16 Path of the annular solar eclipse of 2006 September 22.

unobstructed view toward the east-northeast over the Atlantic, should be especially spectacular. There, annularity will last 5 minutes 42 seconds.

The central-eclipse path then plunges into the Atlantic for a long, watery trip toward the southeast. The path never touches land again, passing several hundred kilometers to the west of the Cape of Good Hope as well as the South Atlantic islands of Saint Helena and Ascension.

Table 7.46 **Annular Solar Eclipse of 2006 September 22**
Center Line Coordinates

U.T. hh:mm	Longitude °	′	Latitude °	′	Alt. °	Path Width km	Duration m:ss
09:53	−50	00	+04	49.1	10	318	5:46
10:01	−40	00	+03	05.2	22	314	6:08
10:16	−30	00	−00	37.4	36	309	6:30
10:46	−20	00	−07	39.2	53	291	6:55

Table 7.46 **(Continued)**

U.T. hh:mm	Longitude °	′	Latitude °	′	Alt. °	Path Width km	Duration m:ss
11:35	−10	00	−19	24.0	66	263	7:08
11:41	**−09**	**03.4**	**−20**	**39.9**	**66**	**261**	**7:09 (GE)**
12:21	00	00	−31	39.6	57	255	7:00
12:50	+10	00	−39	60.0	44	262	6:43
13:06	+20	00	−45	22.0	33	273	6:24
13:16	+30	00	−48	53.8	24	284	6:10
13:23	+40	00	−51	12.0	17	295	5:57
13:26	+50	00	−52	36.1	08	306	5:46
13:28	+60	00	−53	17.5	03	311	5:42

Table 7.47 **Local Contact Times: 2006 September 22**

  a. Annular

City/town	P1	U1	Maximum	U2	P2	Duration m:ss
Cayenne, French Guiana	—	09:49	09:52	09:55	11:10	5:42
St. Laurent, French Guiana	—	09:49	09:51	09:54	11:08	5:29
Paramaribo, Suriname	—	09.48	09:51	09:53	11:06	5:00
Oiapoque, Brazil	—	09:52	09:53	09:55	11:12	3:31

  b. Partial

City	P1	Maximum	P2	Magnitude
Rio de Janeiro	09:41	10:47	12:02	0.395
São Paulo, Brazil	09:42	10:44	11:54	0.358
Cape Town	11:20	12:56	14:24	0.701
Johannesburg	11:43	13:03	14:15	0.381

Table 7.48 **Weather Prospects: 2006 September 22**

City	Clear	Scattered	Broken	Overcast
Cayenne, French Guiana	1.1	75.8	22.4	0.7
Paramaribo, Suriname	4.3	74.9	19.8	0.5

PARTIAL SOLAR ECLIPSE OF 2007 MARCH 19 (SAROS 149)

Eclipse chasers will have a bit of a hiatus this year, with only two partial eclipses occurring in 2007. This first partial event will cover most of central and eastern Asia as well as westernmost Alaska. Nome, for instance, will see a minor eclipse on the afternoon of March 18, the date shift caused by the circumstances of the eclipse relative to the International Date Line.

At its greatest point, lying just west of the Ural Mountains near the Russian village of Timser, the eclipse will reach a magnitude of 0.874 just as the Sun sets.

Table 7.49 **Local Contact Times: 2007 March 19**
(Greatest eclipse: Time 02:33 u.t., Latitude 61° 02.9′, Longitude 55° 24.7′, Mag. 0.874)

City	P1	Maximum	P2	Magnitude
Beijing	01:27	02:23	03:21	0.395
Delhi	—	01:37	02:31	0.577
Nome, Alaska	03:25	03:51	04:17	0.103
Seoul	01:47	02:31	03:15	0.195

PARTIAL SOLAR ECLIPSE OF 2007 SEPTEMBER 11 (SAROS 154)

The Moon's penumbral shadow passes over a portion of South America during this southern hemisphere, late-winter partial eclipse. South of a cutoff line that passes through central Peru and Brazil, observers will see the Moon slowly slip across the solar disk during the morning hours. Anyone along the continent's west coast will see an already eclipsed Sun rise above the eastern horizon; others in the interior and along the east coast will see the eclipse's full span. Greatest eclipse, at magnitude 0.749, occurs in the far South Pacific, hundreds of kilometers offshore from Cape Horn.

Table 7.50 **Local Contact Times: 2007 September 11**
(Greatest eclipse: Time: 12:32 u.t., Latitude −61° 00.3′, Longitude −90° 17.4, Mag. 0.749)

City	P1	Maximum	P2	Magnitude
Asunción, Paraguay	10:31	11:30	12:35	0.389
Rio de Janeiro	10:43	11:39	12:39	0.257
Santiago, Chile	—	11.39	12:48	0.550
Buenos Aires	10:41	11:48	13:01	0.519
Stanley, Falkland Islands	11:13	12:25	13:42	0.653
Cape Horn, Chile	11:18	12:27	13:42	0.709

ANNULAR SOLAR ECLIPSE OF 2008 FEBRUARY 7 (SAROS 121)

Once again we are enticed by an annular eclipse that few if any of us will see. More than half of the maximum-eclipse track occurs over water, and of the portion that does pass over dry land, that land is Antarctica. As seen in figure 7.17, the only inhabited site that will see annularity is the Russian scientific research station Russkaya, where annularity will last 2 minutes 8 seconds.

The land areas witnessing a partial eclipse will also be greatly limited. Only observers in New Zealand, southeastern Australia, and a few South Pacific islands will see any effect.

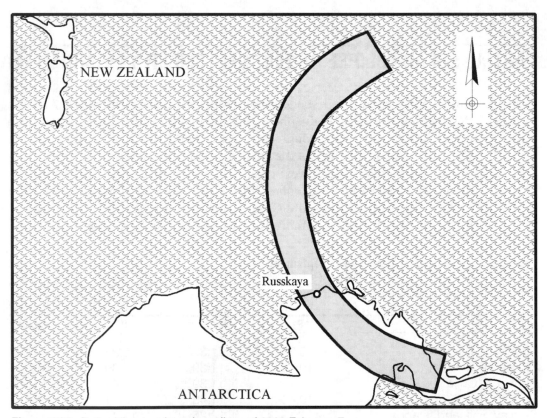

Figure 7.17   Path of the annular solar eclipse of 2008 February 7.

Table 7.51   **Annular Solar Eclipse of 2008 February 7**
Center Line Coordinates

U.T. hh:mm	Longitude °	 ′	Latitude °	 ′	Alt. °	Path Width km	Duration m:ss
03:24	−80	00	−75	50.9	03	537	2:14
03:26	−90	00	−77	32.8	06	518	2:14
03:28	−100	00	−78	16.0	08	494	2:13
03:30	−110	00	−78	23.1	10	478	2:13
03:32	−120	00	−77	57.7	11	461	2:13
03:35	−130	00	−76	51.8	13	448	2:13
03:41	−140	00	−74	34.7	14	438	2:13
03:53	−150	00	−68	11.3	16	441	2:12
03:56	**−150**	**26.9**	**−67**	**34.6**	**16**	**445**	**2:12 (GE)**
04:13	−150	00	−57	51.1	13	512	2:12
04:25	−140	00	−49	31.0	04	572	2:12

Table 7.52   **Local Contact Times: 2008 February 7**
a. Annular

City/town	P1	U1	Maximum	U2	P2	Duration m:ss
Russkaya Station	02:39	03:39	03:40	03:41	04:40	2:08

b. Partial

City	P1	Maximum	P2	Magnitude
Auckland, New Zealand	03:46	04:51	05:51	0.574
Canberra, Australia	03:47	04:38	05:26	0.206
Christchurch, New Zealand	03:28	04:37	05:39	0.627
Melbourne	03:38	04:28	05:15	0.183
Sydney	03:52	04:43	05:32	0.217

TOTAL SOLAR ECLIPSE OF 2008 AUGUST 1 (SAROS 126)

Umbraphiles, having endured nearly two and a half years of withdrawal, will no doubt rejoice at the coming of this, the first total solar eclipse since March 2006. And while the path of totality hugs some of Earth's remotest regions (figure 7.18), eclipse chasers will still likely globe-trot just to get their two-minute "fix."

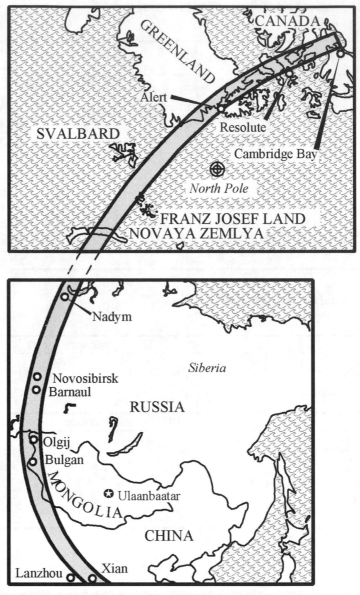

Figure 7.18 Track of the total solar eclipse of 2008 August 1.

The Moon's umbra is first cast upon the Earth in the far northern corner of the Northwest Territories, Canada. Though the central shadow narrowly misses the isolated towns of Cambridge Bay, on Victoria Island, and Resolute, on Cornwallis Island, its edge just nips the town of Alert, on Ellesmere Island, giving residents 40 early-morning seconds of totality.

Crossing the frigid Arctic north, the path skims across the northernmost coast of Greenland and comes within about 720 kilometers (450 miles) of the North Pole before heading southward toward more-moderate climes. Totality passes by the Norwegian island group of Svalbard, just touches Russia's Franz Josef Land island group, and cuts across the crescent-shaped island of Novaya Zemlya on its way to mainland Asia. The umbra first touches the Russian coast on the Yamal Peninsula. Not far inland, greatest eclipse, producing 2 minutes 27 seconds of totality, is reached near the tiny town of Nadym, just inland from the boot-shaped Gulf of Obskaja.

Continuing to hook toward the southeast, the central path passes very near the city of Novosibirsk, where totality lasts 2 minutes 18 seconds. The path then briefly enters western Mongolia, with the towns of Olgij and Bulgan seeing about 2 minutes of totality. Totality finally whisks into north-central China before leaving Earth at a point just north of the cities of Lanzhou and Xian.

Table 7.53 **Total Solar Eclipse of 2008 August 1**

Center Line Coordinates

U.T. hh:mm	Longitude °	′	Latitude °	′	Alt. °	Path Width km	Duration m:ss
09:22	−100	00	+69	54.7	02	205	1:30
09:24	−90	00	+74	31.4	08	214	1:43
09:26	−80	00	+77	50.6	12	217	1:49
09:28	−70	00	+80	04.5	14	218	1:53
09:30	−60	00	+81	33.1	16	219	1:57
09:32	−50	00	+82	31.7	18	220	2:00
09:33	−40	00	+83	09.7	19	220	2:00
09:35	−30	00	+83	32.8	20	221	2:04
09:36	−20	00	+83	44.1	21	221	2:05
09:38	−10	00	+83	45.0	22	222	2:06
09:39	00	00	+83	35.5	23	222	2:09
09:41	+10	00	+83	14.5	24	223	2:11
09:43	+20	00	+82	38.7	25	223	2:13
09:45	+30	00	+81	42.6	26	224	2:14
09:49	+40	00	+80	15.3	28	224	2:17
09:54	+50	00	+77	56.4	29	226	2:21
10:02	+60	00	+74	06.2	32	229	2:24
10:16	+70	00	+67	37.1	33	234	2:27
10:22	**+72**	**16.3**	**+65**	**38.2**	**34**	**236**	**2:27 (GE)**
10:38	+80	00	+57	50.4	32	244	2:23
11:00	+90	00	+47	34.0	24	251	2:08
11:14	+100	00	+39	55.3	14	236	1:49
11:19	+110	00	+34	53.6	04	210	1:33

Table 7.54 **Local Contact Times: 2008 August 1**

a. Total

City/town	P1	U1	Maximum	U2	P2	Duration m:ss
Alert, Northwest Territories	08:36	09:32	09:32	09:32	10:29	0:40
Nadym, Russia	09:16	10:20	10:21	10:22	11:23	2:26
Novosibirsk, Russia	09:41	10:43	10:45	10:46	11:45	2:18
Barnaul, Russia	09:44	10:47	10:48	10:49	11:48	2:16
Olgij, Mongolia	09:57	10:58	10:59	11:00	11:56	1:56
Bulgan, Mongolia	10:01	11:01	11:02	11:03	11:59	2:04

b. Partial

City	P1	Maximum	P2	Magnitude
Berlin	08:43	09:37	10:32	0.300
Delhi	10:32	11:31	12:26	0.628
Lanzhou, China	10:23	11:19	—	0.992
London	08:32	09:17	10:04	0.218
Moscow	09:01	10:09	11:14	0.579
Paris	08:42	09:20	10:00	0.143
St. John's, Newfoundland	—	08:43	09:24	0.317

Table 7.55 **Weather Prospects: 2008 August 1**

City	Clear	Scattered	Broken	Overcast
Alert, Northwest Territories	2.3	19.1	31.3	41.1
Barnaul, Russiaa	Clouds < $3/10$ = 5.5 days		Precip > 0.1 inch = 8.3 days	
Bulgan, Mongolia	Clouds < $3/10$ = 7.8 days		Precip > 0.1 inch = 7.4 days	
Novosibirsk, Russia	Clouds < $3/10$ = 4.9 days		Precip > 0.1 inch = 9.9 days	
Resolute, Canada	1.4	14.6	31.9	41.9

ANNULAR SOLAR ECLIPSE OF 2009 JANUARY 26 (SAROS 131)

The year 2009 brings with it a pair of watery solar eclipses, beginning with this annular event. The track of annularity first meets Earth in the South Atlantic, dodging land as it passes hundreds of kilometers south of Cape Horn and the African continent. The point of greatest eclipse, with 7 minutes 54 seconds of annularity, occurs in the Indian Ocean, about halfway between Madagascar and Australia.

Annularity finally strikes land in Indonesia just before sunset. The southern portion of Sumatra, including the mighty volcano Krakatoa, off the island nation's southern tip, and northernmost Java will witness the annular phase firsthand, with the Sun a little more than 20° above the western horizon. The cities of Kotabumi and Telukbetung, each located within a few kilometers of the exact center line, will witness more than 6 minutes of annularity, while Krakatoa, closer to the shadow's edge, will see a little less than 5 minutes' worth. The central path will then hop across the Karimata Strait to land on Borneo. The cities of Sampit and Samarinda, located near the southern extreme of annularity, will each witness a lopsided, ring-of-fire sunset eclipse (figure 7.19).

Table 7.56 **Annular Solar Eclipse of 2009 January 26**
Center Line Coordinates

U.T. hh:mm	Longitude °	′	Latitude °	′	Alt. °	Path Width km	Duration m:ss
06:05	−10	00	−35	09.9	+01	360	5:44
06:07	+00	00	−38	20.3	+11	347	6:02
06:12	+10	00	−40	55.3	+20	332	6:20
06:20	+20	00	−42	45.9	+29	318	6:40
06:31	+30	00	−43	44.3	+38	305	7:01
06:45	+40	00	−43	41.5	+47	294	7:22
07:03	+50	00	−42	24.0	+57	286	7:41
07:27	+60	00	−39	28.2	+67	280	7:54
07:57	+70	00	−34	15.5	+73	280	7:54
07:59	**+70**	**16.4**	**−34**	**04.7**	**+73**	**280**	**7:54 (GE)**
08:34	+80	00	−26	14.3	+65	292	7:34
09:10	+90	00	−16	45.1	+48	316	6:59
09:34	+100	00	−08	30.3	+31	336	6:27
09:46	+110	00	−02	18.0	+16	349	6:03
09:51	+120	00	+02	14.3	+04	358	5:44

Table 7.57 **Local Contact Times: 2009 January 26**
a. Annular

City/town	P1	U1	Maximum	U2	P2	Duration m:ss
Krakatoa	08:18	09:38	09:40	09:42	10:50	4:43
Serang, Java	08:19	09:39	09:40	09:41	10:50	2:39
Telukbetung, Sumatra	08:19	09:38	09:41	09:44	10:51	6:07
Kotabumi, Sumatra	08:20	09:38	09:41	09:44	10:51	6:09
Sampit, Borneo	08:31	09:45	09:46	—	10:52	1:24
Samarinda, Borneo	08:36	09:47	09:48	09:50	—	2:27

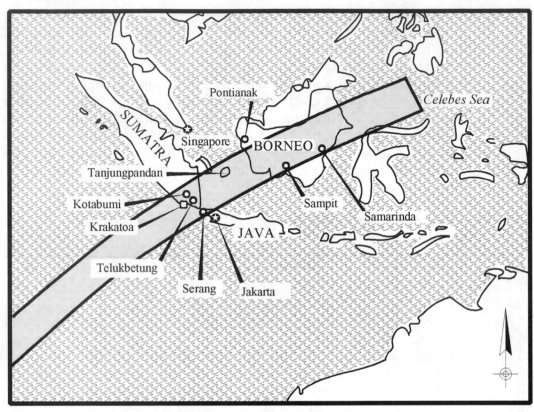

Figure 7.19 Path of the annular solar eclipse of 2009 January 26.

Table 7.57 **(Continued)**

> *b. Partial*

City	P1	Maximum	P2	Magnitude
Cape Town	04:58	06:11	07:37	0.720
Jakarta, Java	08:20	09:40	10:50	0.914
Johannesburg	05:06	06:19	07:45	0.461
Perth, Australia	08:01	09:01	09:55	0.336
Pretoria	05:06	06:19	07:45	0.449
Singapore	08:29	09:49	10:57	0.796

Table 7.58 **Weather Prospects: 2009 January 26**

City	Clear	Scattered	Broken	Overcast
Jakarta, Java	0	4.7	71.2	24.1
Pontianak, Borneo	0	6.6	67.5	26.0
Tanjungpandan, Belitung	0	4.4	77.7	17.9

TOTAL SOLAR ECLIPSE OF 2009 JULY 22 (SAROS 136)

Saros 136 is well known to umbraphiles for producing the longest-lasting total eclipses of the twentieth century. The previous eclipse in the Saros occurred on 1991 July 10, when totality crossed Hawaii and portions of Mexico, and Central and South America. In 2009, Asia is the scene for the longest total eclipse to occur within the twenty-year span of this book. The umbra first touches down along the west coast of the Indian subcontinent before following a northeasterly track (figure 7.20). Just nipping Nepal's east border and the western tip of Bangladesh, the umbra completely engulfs the tiny nation of Bhutan before entering China.

Totality passes over many cities and towns within south-central China, including Chengdu, Chongqing (Chunking), and Shanghai. A favorite gathering point for many eclipse chasers is likely the city of Wuhan, where the total phase lasts more than five minutes, longer than any other major city along the path.

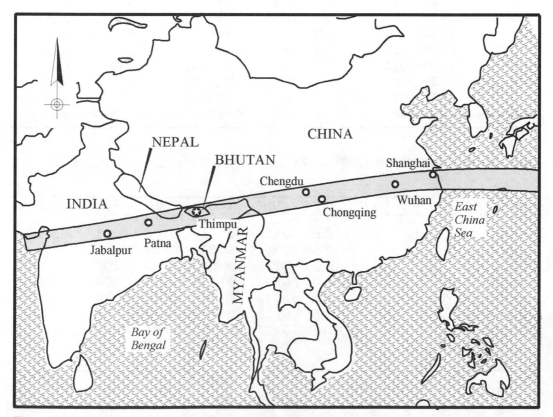

Figure 7.20   Track of the total solar eclipse of 2009 July 22.

The shadow then leaves dry land for the broad scope of the Pacific Ocean, just nipping the tiny Japanese island of Amami, but missing the nation's main islands. Greatest eclipse, lasting an enticingly long 6 minutes 39 seconds, occurs at a watery point several hundred kilometers northeast of Iwo Jima. The eclipse finally draws to a close as the central shadow leaves Earth at sunset far to the south of Hawaii.

Table 7.59 **Total Solar Eclipse of 2009 July 22**
Center Line Coordinates

U.T. hh:mm	Longitude °	′	Latitude °	′	Alt. °	Path Width km	Duration m:ss
00:54	+80	00	+23	43.7	+10	216	3:37
00:59	+90	00	+26	48.0	+20	224	4:03
01:07	+100	00	+29	08.6	+31	232	4:37
01:20	+110	00	+30	31.8	+43	241	5:14
01:36	+120	00	+30	43.2	+55	248	5:50
01:57	+130	00	+29	24.6	+68	254	6:21
02:23	+140	00	+26	11.9	+82	258	6:39
02:36	**+144**	**08.4**	**24**	**12.0**	**+86**	**258**	**6:39 (GE)**
02:53	+150	00	+20	42.3	+79	258	6:30
03:24	+160	00	+13	12.8	+61	255	5:52
03:50	+170	00	+05	13.4	+43	247	5:02
04:06	180	00	−01	49.6	+27	235	4:17
04:14	−170	00	−07	32.0	+14	221	3:41
04:17	−160	00	−12	01.5	+03	205	3:09

Table 7.60 **Local Contact Times: 2009 July 22**
a. Total

City/town	P1	U1	Maximum	U2	P2	Duration m:ss
Jabalpur, India	23:59	00:52	00:53	00:55	01:54	3:02
Patna, India	23:59	00:54	00:56	00:58	01:59	3:46
Thimpu, Bhutan	00:00	00:58	00:59	01:00	02:04	2:39
Chengdu, China	00:06	01:11	01:12	01:14	02:26	3:21
Chongqing, China	00:07	01:13	01:15	01:17	02:30	3:55
Wuhan, China	00:14	01:23	01:26	01:29	02:46	5:25
Shanghai, China	00:23	01:37	01:39	01:41	03:01	3:37

b. Partial

City	P1	Maximum	P2	Magnitude
Beijing	00:25	01:32	02:44	0.730
Hilo, Hawaii	03:24	03:50	04:15	0.117

Table 7.60 **(Continued)**

City	P1	Maximum	P2	Magnitude
Honolulu	03:21	03:47	04:13	0.116
Kyoto	00:47	02:05	03:25	0.809
Osaka	00:46	02:05	03:25	0.822
Tokyo	00:55	02:12	03:30	0.748

Table 7.61 **Weather Prospects: 2009 July 22**

City	Clear	Scattered	Broken	Overcast
Chengdu, China	4.3	12.5	33.0	48.9
Chongqing, China	5.6	20.2	38.7	32.8
Jabalpur, India	0.2	7.0	30.3	62.5
Patna, India	0	10.1	63.6	26.1
Shanghai	3.0	20.9	50.1	26.1
Wuhan, China	11.2	23.3	43.0	22.0

ANNULAR SOLAR ECLIPSE OF 2010 JANUARY 15 (SAROS 141)

Here's another event for the twenty-first-century record books. At its greatest, the annular phase of this eclipse will last an incredible 11 minutes 8 seconds, the longest since 1992 January 4. But what is even more impressive is that this duration will not be equaled or exceeded until the annular eclipse of 3043 December 23 (although there will be 21 "10-plus-minute" annular eclipses between now and then)!

Unfortunately, the maximum-eclipse point occurs in the middle of the Indian Ocean, but we landlubbers will still have ample opportunity to see a somewhat abbreviated annular phase from either Africa or Asia (figure 7.21). Annularity will greet observers in the towns of Bambari and Bangassou, Central African Republic, at sunrise, then quickly head east into Uganda. There, the capital city of Kampala will witness 7 minutes 39 seconds of annularity, although the Sun will only be 20° above the eastern horizon at the time.

Skimming across the north shore of Lake Victoria, annularity will next pass into Kenya. There, the city of Nakuru will be bathed in more than 8 minutes of annularity, while from the capital city of Nairobi, set toward the central path's southern edge, the annular phase will last just under 6 minutes. Beginning to hook toward the northeast, annularity will quickly cross

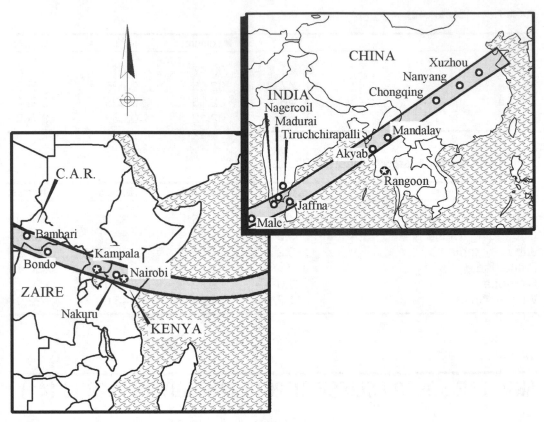

Figure 7.21 Path of the annular solar eclipse of 2010 January 15.

southernmost Somalia before leaving Africa for a cruise across the Indian Ocean.

The path will next land on the tiny island nation of Maldives, where annularity will last an astonishing 10 minutes 44 seconds. Then, it is on to India, where towns along the path's center, such as Nagercoil, will see nearly 10 minutes of annularity. The northern half of the island nation of Sri Lanka will also see the annular phase, with the city of Jaffna being treated to more than 10 minutes of annularity.

Crossing the Bay of Bengal, annularity will sweep briefly over southeastern Bangladesh before moving on into Myanmar, where it will pass over the towns of Akyab and Mandalay. The path will make a final country-hop into China, where it will bring to Nanyang 7 minutes 26 seconds of annularity. Xuzhou will be treated to 6 minutes 56 seconds of annularity, with the Sun, by then, getting low in the western sky. But the bonus winners have to be the townsfolk of Chongqing. Only six months ago their city saw a total solar eclipse; now they will again witness the majesty of a central eclipse, with an annular phase lasting 7 minutes 50 seconds.

Table 7.62 **Annular Solar Eclipse of 2010 January 15**
Center Line Coordinates

U.T. hh:mm	Longitude °	′	Latitude °	′	Alt. °	Path Width km	Duration m:ss
05:17	+20	00	+05	18.2	+04	368	7:15
05:22	+30	00	+01	44.4	+16	356	8:00
05:34	+40	00	−01	07.8	+29	349	8:49
05:54	+50	00	−02	43.3	+44	346	9:52
06:26	+60	00	−02	05.3	+59	344	10:54
07:07	**+69**	**20.2**	**+01**	**37.3**	**+66**	**333**	**11:08 (GE)**
07:09	+70	00	+02	01.2	+66	332	11:06
07:54	+80	00	+09	40.5	+55	323	10:10
08:26	+90	00	+18	11.1	+38	330	9:03
08:44	+100	00	+25	30.2	+24	343	8:13
08:52	+110	00	+31	23.8	+12	357	7:39
08:55	+120	00	+36	06.6	+02	372	7:12

Table 7.63 **Local Contact Times: 2010 January 15**
a. Annular

City/town	P1	U1	Maximum	U2	P2	Duration m:ss
Bambari, C.A.R.	—	05:14	05:18	05:21	06:43	6:49
Bondo, Zaire	—	05:15	05:18	05:22	06:46	7:33
Kampala, Uganda	04:05	05:21	05:24	05:28	07:03	7:39
Nakuru, Kenya	04:06	05:24	05:28	05:33	07:13	8:25
Nairobi, Kenya	04:06	05:26	05:29	05:32	07:14	6:53
Male, Maldives	05:14	07:19	07:24	07:30	09:22	10:44
Nagercoil, India	05:35	07:40	07:44	07:49	09:35	9:47
Madurai, India	05:41	07:48	07:49	07:51	09:38	2:47
Jaffna, Sri Lanka	05:47	07:49	07:54	07:59	09:41	10:10
Akyab, Myanmar	06:45	08:28	08:32	08:37	10:01	8:39
Mandalay, Myanmar	06:55	08:34	08:38	08:42	10:03	7:38
Chongqing, China	07:20	08:46	08:50	08:54	10:07	7:50
Nanyang, China	07:29	08:49	08:53	08:57	10:06	7:26
Xuzhou, China	07:34	08:51	08:55	08:58	10:05	6:56

b. Partial

City	P1	Maximum	P2	Magnitude
Bombay	05:46	07:47	09:34	0.644
Calcutta	06:37	08:27	09:58	0.835

Table 7.63 **(Continued)**

City	P1	Maximum	P2	Magnitude
Delhi	06:23	08:09	09:40	0.526
Johannesburg	04:41	05:25	06:14	0.177
Pretoria	04:39	05:25	06:16	0.189
Singapore	07:09	08:26	09:32	0.333

Table 7.64 **Weather Prospects: 2010 January 15**

City	Clear	Scattered	Broken	Overcast
Chongqing, China	17.5	8.3	13.3	60.8
Nairobi, Kenya	3.0	40.6	50.6	5.8
Rangoon, Myanmar	26.9	54.6	16.4	2.1
Tiruchchirapalli, India	2.6	53.4	40.9	3.1
Xuzhou, China	30.1	15.5	27.6	26.8

TOTAL SOLAR ECLIPSE OF 2010 JULY 11 (SAROS 146)

Here we have yet another eclipse made for fish! The track of this total event spans the width of the South Pacific, yet, apart from passing over a lone tiny island, never touches land until sunset.

The Moon's umbral shadow first touches down in the Pacific Ocean nearly 1,600 kilometers (1,000 miles) northeast of New Zealand. It then heads east-northeast in a long arc that takes it north of Pitcairn Island. The point of greatest eclipse, producing 5 minutes 20 seconds of totality, occurs far from land, about halfway between New Zealand and South America.

Hooking southeast, the umbra crosses Easter Island, annexed by Chile and nestled some 3,700 kilometers (about 2,300 miles) west of the Chilean coast. Easter Island is best known as the home of strange humanoid megalithic statues, the exact origin of which remains one of history's greatest archaeological mysteries. Talk about a thrilling photo opportunity! Eclipse chasers will have the unprecedented chance to capture the megaliths silhouetted in front of the "angry" solar eclipse. And, to add further enticement, Easter Island will experience 4 minutes 44 seconds of totality, the longest of any landfall (although clouds may be a problem; see below).

The umbra then speeds southeastward on its way toward South America, where, as sunset nears, it just tickles the southern portions of Chile and

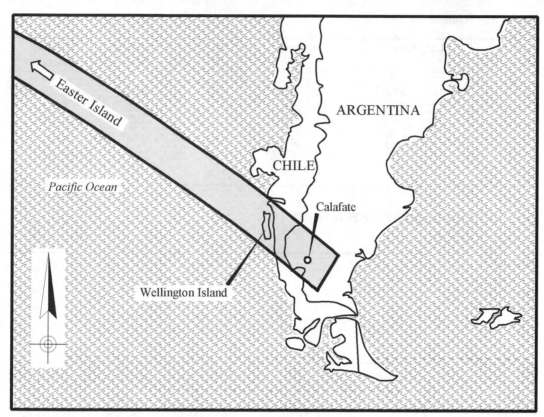

Figure 7.22   Track of the total solar eclipse of 2010 July 11.

Argentina (figure 7.22). Wellington Island, off the jagged, island-strewn coast of Chile, experiences 2 minutes 46 seconds of totality, with the Sun just a few degrees above the horizon. Farther inland, the town of Calafate, Argentina, sees 2 minutes 47 seconds, with the Sun setting just as totality ends. In both cases, a near-perfect horizon is a must.

Table 7.65   **Total Solar Eclipse of 2010 July 11**
Center Line Coordinates

U.T. hh:mm	Longitude °	′	Latitude °	′	Alt. °	Path Width km	Duration m:ss
18:16	−170	00	−26	28.3	+01	179	2:42
18:19	−160	00	−22	33.0	+11	196	3:14
18:28	−150	00	−19	18.8	+23	216	3:50
18:44	−140	00	−17	20.3	+34	240	4:32
19:08	−130	00	−17	28.7	+44	259	5:09

Table 7.65 **(Continued)**

U.T. hh:mm	Longitude °	′	Latitude °	′	Alt. °	Path Width km	Duration m:ss
19:34	**−121**	**51.0**	**−19**	**46.5**	**+47**	**259**	**5:20 (GE)**
19:39	−120	00	−20	34.2	+47	255	5:18
20:08	−110	00	−26	26.6	+40	233	4:51
20:30	−100	00	−33	30.3	+29	214	4:10
20:42	−90	00	−40	17.7	+18	199	3:33
20:48	−80	00	−46	15.0	+08	189	3:04

Table 7.66 **Local Contact Times: 2010 July 11**
          a. Total

City/town	P1	U1	Maximum	U2	P2	Duration m:ss
Easter Island	18:40	20:08	20:10	20:13	21:34	4:44
Wellington Is., Chile	19:42	20:48	20:49	20:50	—	2:46
Calafate, Argentina	19:44	20:48	20:49	—	—	2:47

b. Partial

City	P1	Maximum	P2	Magnitude
Asunción, Paraguay	20:33	21:06	—	0.173
La Paz, Bolivia	20:39	21:01	21:22	0.061
Lima, Peru	20:24	20:52	21:18	0.080
Santiago, Chile	20:00	21:00	—	0.583

Table 7.67 **Weather Prospects: 2010 July 11**

City	Clear	Scattered	Broken	Overcast
Calafate, Argentina	Clouds < $^3/_{10}$ = 5.8 days		Precip > 0.1 inch = 1.2 days	
Easter Island	0.6	29.2	49.9	20.3

PARTIAL SOLAR ECLIPSE OF 2011 JANUARY 4 (SAROS 151)

The year 2011 will be jam-packed with four solar eclipses, although all will be partial events. This first eclipse covers nearly all of Europe, the northern half of Africa, the Middle East, and southern Asia, affording many the opportunity to view at least a small portion of the Sun's northern limb blocked by the Moon. Greatest eclipse, reaching magnitude 0.858, occurs at sunrise in northeastern Sweden. Towns and cities in western Europe, including London and Paris, also enjoy a sunrise eclipse (such as that captured in Figure 7.23), while the northerly latitude of Stockholm, Sweden, causes the thin-crescent Sun to move slowly along the horizon.

Table 7.68 **Local Contact Times: 2011 January 4**
(Greatest eclipse: Time 08:51 u.t., Latitude 64° 39.1', Longitude 20° 48.8', Mag. 0.858)

City	P1	Maximum	P2	Magnitude
Cairo	07:01	08:30	10:05	0.551
Jerusalem	07:10	08:41	10:16	0.568
London	—	—	09:30	0.748
Moscow	07:37	09:03	10:29	0.812
Paris	—	08:09	09:30	0.732
Rome	06:51	08:10	09:38	0.696
Stockholm	—	08:41	10:04	0.849

Figure 7.23 Ernest W. Piini took this magnificent photograph of the rising Sun in partial eclipse from South Mountain, near Phoenix, Arizona. (88-mm f/6.8 refractor stopped down to f/22, 1/125th-second exposure on Kodachrome II ISO 25 film.)

PARTIAL SOLAR ECLIPSE OF 2011 JUNE 1 (SAROS 118)

This second partial eclipse of 2011 will be enjoyed only by amateurs who find themselves traveling north for the summer. There, observers in northern Alaska, northern Canada, Greenland, Iceland, and portions of northeastern Asia will see a relatively minor partial eclipse sometime during the day. Greatest eclipse, at magnitude 0.602, will be seen in far northwestern Russia, near Cheshskaya Bay. The cities mentioned below, however, will experience a much less substantial event.

Table 7.69 **Local Contact Times: 2011 June 1**
(Greatest eclipse: Time 21:17 u.t., Latitude 67° 47.0', Longitude 46° 49.4', Mag. 0.602)

City	P1	Maximum	P2	Magnitude
Barrow, Alaska	20:22	21:05	21:49	0.187
Fairbanks	20:47	21:09	21:30	0.034
Nome, Alaska	20:19	20:49	21:20	0.080
Sapporo, Japan	19:26	19:50	20:14	0.086
St. John's, Newfoundland	22:11	22:39	23:06	0.124

PARTIAL SOLAR ECLIPSE OF 2011 JULY 1 (SAROS 156)

Just one lunation later, a third partial eclipse will occur, though the odds are good that no one will see it. This time, the brief touch of the lunar penumbra will only glance the globe off Lutzow-Holm Bay on the coast of Antarctica, near where the South Atlantic and Indian Oceans meet. Even at its greatest, this eclipse will only reach a magnitude of 0.096. Greatest eclipse will be achieved at 08:39 u.t., at latitude −65° 09.5', longitude +28° 38.9'.

PARTIAL SOLAR ECLIPSE OF 2011 NOVEMBER 25 (SAROS 123)

This final partial eclipse of 2011 will also be visible from a limited area of the southern hemisphere. The penumbra will again be centered near

Antarctica, while its outermost edge will pass over portions of New Zealand's South Island, Tasmania, and southernmost South Africa. The eclipse's greatest point, at magnitude 0.904, will again remain inaccessibly near the coast of Antarctica.

Table 7.70 **Local Contact Times: 2011 November 25**
(Greatest eclipse: Time 06:21 u.t., Latitude −68° 34.1′, Longitude −82° 24.0′, Mag. 0.904)

City	P1	Maximum	P2	Magnitude
Cape Town	04:28	04:52	05:17	0.102
Dunedin, New Zealand	07:03	07:40	—	0.305
Hobart, Tasmania	07:30	07:49	08:07	0.055

ANNULAR SOLAR ECLIPSE OF 2012 MAY 20–21 (SAROS 128)

Although 2011 will leave eclipse chasers wanting, 2012 should prove more satisfying. The year will kick off with a widely visible annular eclipse spanning the Pacific rim and crossing a portion of the western United States (figure 7.24). This will be the first time an annular eclipse is visible in the United States since 1994 May 10. This may help satisfy some Americans' thirst, but we still have to wait five more years for a total solar eclipse to grace our land.

Beginning at sunrise along the Chinese coast, the zone of annularity crosses Hong Kong and Guangzhou (formerly known as Canton), where millions of people have a chance to see annularity (lasting 3 minutes 30 seconds and 4 minutes 25 seconds, respectively). A short hop over the East China Sea brings annularity to northern Taiwan and southeastern Japan. On Taiwan, residents in and around Taipei enjoy a comparatively short 2 minutes 2 seconds of annularity, owing to their placement near the southern boundary of the track of annularity. Again, millions are thrust into annularity as the shadow crosses over a number of towns and cities including Tokyo, where residents witness the longest annular phase of any major city along the track, 5 minutes 4 seconds.

A dive into the Pacific sets the path on an eastbound trans-oceanic cruise. The point of greatest eclipse, where annularity lasts 5 minutes 46 seconds, takes place to the south of Kiska and Buldir, in Alaska's Aleutian Island chain.

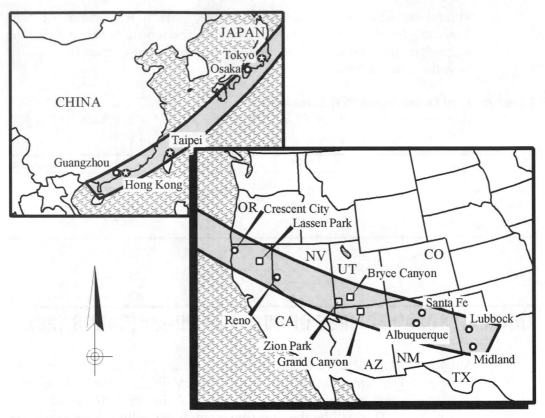

Figure 7.24 Path of the annular solar eclipse of 2012 May 20.

Annularity finally comes ashore near Crescent City, California, and proceeds through portions of Nevada, Utah, Arizona, Colorado, and New Mexico before leaving Earth in Texas at sunset. Some of the cities along the path include Reno, Nevada; Santa Fe and Albuquerque, New Mexico; and Lubbock and Midland, Texas. Even more exciting, the path crosses several national parks, including Lassen Volcanic National Park in California, Zion National Park and Bryce Canyon National Park in Utah, and the Grand Canyon in Arizona. In all four cases the Sun is low in the west, giving photographers a wonderful opportunity to capture both celestial and terrestrial beauty on the same frames of film.

Much of the western United States and Canada witness a partial eclipse this afternoon. The Moon's widespread penumbra reaches as far east as western New York and as far south as Mississippi, although only a small bite of the Sun can be seen along that line by sunset.

Table 7.71 **Annular Solar Eclipse of 2012 May 20–21**
Center Line Coordinates

U.T. hh:mm	Longitude °	′	Latitude °	′	Alt. °	Path Width km	Duration m:ss
22:08	+110	00	+21	37.0	+01	324	4:23
22:11	+120	00	+25	46.7	+12	313	4:35
22:20	+130	00	+30	33.9	+23	298	4:49
22:34	+140	00	+35	43.5	+35	279	5:05
22:54	+150	00	+40	42.2	+46	261	5:22
23:16	+160	00	+44	51.6	+55	247	5:36
23:39	+170	00	+47	50.0	+60	238	5:45
23:53	**+176**	**19.1**	**+49**	**04.6**	**+61**	**237**	**5:46 (GE)**
00:00	180		+49	35.9	+61	237	5:46
00:19	−170	00	+50	15.4	+57	238	5:43
00:37	−160	00	+49	54.4	+52	244	5:34
00:54	−150	00	+48	37.1	+45	254	5:22
01:08	−140	00	+46	27.8	+36	267	5:09
01:20	−130	00	+43	33.3	+27	281	4:55
01:29	−120	00	+40	05.2	+18	298	4:42
01:34	−110	00	+36	18.6	+08	312	4:30

Table 7.72 **Local Contact Times: 2012 May 20–21**
a. Annular

City/town	P1	U1	Maximum	U2	P2	Duration m:ss
Taipei	21:07	22:09	22:10	22:11	23:23	2:02
Hong Kong	21:08	22:06	22:08	22:10	23:16	3:30
Guangzhou, China	21:09	22:07	22:09	22:11	23:17	4:25
Osaka	21:17	22:28	22:29	22:31	23:54	2:44
Tokyo	21:18	22:31	22:34	22:36	00:02	5:04
Crescent City, California	00:07	01:23	01:26	01:28	02:35	4:45
Lassen National Park, California	00:12	01:26	01:28	01:31	02:36	4:43
Reno, Nevada	00:15	01:28	01:30	01:32	02:37	4:26
Carson City, Nevada	00:15	01:29	01:31	01:33	02:37	3:53
Zion National Park, Utah	00:23	01:31	01:34	01:36	—	4:28
Bryce Canyon National Park, Utah	00:22	01:31	01:33	01:35	02:36	4:26
Grand Canyon National Park, Arizona	00:24	01:33	01:35	01:36	02:38	2:51
Santa Fe, New Mexico	00:27	01:32	01:35	01:37	—	4:09
Albuquerque, New Mexico	00:28	01:33	01:35	01:37	—	4:26
Lubbock, Texas	00:30	01:33	01:35	—	—	4:14
Midland, Texas	00:32	01:36	—	—	—	2:38

Table 7.72 **(Continued)**
   b. Partial

City	P1	Maximum	P2	Magnitude
Anchorage	23:16	00:37	01:53	0.678
Beijing	21:31	22:33	23:41	0.670
Denver	00:22	01:29	—	0.855
Juneau, Alaska	23:36	00:54	02:05	0.682
Kyoto	21:17	22:30	23:55	0.940
Los Angeles	00:24	01:38	02:42	0.848
Minneapolis	00:18	01:19	—	0.671
Phoenix	00:28	01:38	—	0.887
Portland, Oregon	00:03	01:20	02:29	0.873
San Francisco	00:15	01:32	02:39	0.898
Seattle	00:01	01:17	02:25	0.828
Vancouver	23:58	01:14	02:23	0.800

Table 7.73 **Weather Prospects: 2012 May 20–21**

City	Clear	Scattered	Broken	Overcast
Albuquerque, New Mexico	31.1	26.4	24.7	17.8
Flagstaff, Arizona	25.2	29.5	30.7	14.6
Guangzhou, China	6.2	8.2	40.4	45.1
Redding, California	22.6	28.5	29.6	19.4
Reno, Nevada	18.3	27.0	26.8	27.9
Taipei, Taiwan	1.4	20.2	49.1	29.0
Yokohama/Tokyo, Japan	7.6	19.2	35.5	37.8

TOTAL SOLAR ECLIPSE OF 2012 NOVEMBER 13 (SAROS 133)

Like the first solar eclipse of 2012, this total event also travels west to east across the Pacific Ocean (figure 7.25), but following a more southerly track that nearly avoids land altogether. The largest population center to see totality is the city of Cairns, on the eastern shore of Australia's Cape York Peninsula. There the Sun rises at first contact and is only 13° above the horizon when totality begins.

The umbra then quickly sails out into the South Pacific, following a fine thread that takes totality south of New Caledonia and north of Norfolk

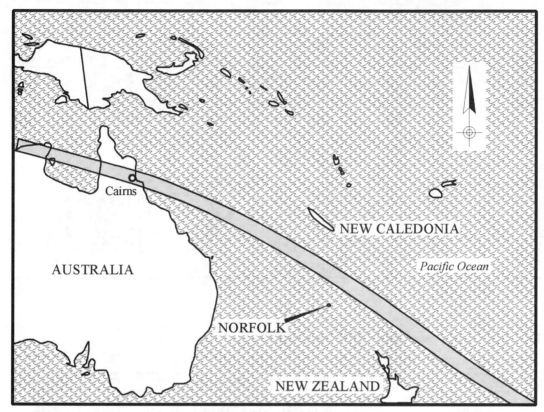

Figure 7.25 Track of the total solar eclipse of 2012 November 13.

Island and New Zealand. Greatest eclipse, where totality lasts 4 minutes 2 seconds, occurs far out at sea, some 1,800 kilometers (roughly 1,100 miles) east of New Zealand's North Island. The path continues eastward toward the Chilean coast, but leaves the Earth at sunset still more than 800 kilometers (500 miles) offshore.

Table 7.74 **Total Solar Eclipse of 2012 November 13**
Center Line Coordinates

U.T. hh:mm	Longitude °	′	Latitude °	′	Alt. °	Path Width km	Duration m:ss
20:36	+140	00	−14	24.5	+07	136	1:54
20:41	+150	00	−18	32.3	+19	150	2:15
20:52	+160	00	−23	17.6	+31	163	2:42
21:08	+170	00	−28	24.2	+44	172	3:11

Table 7.74 **(Continued)**

U.T. hh:mm	Longitude °	′	Latitude °	′	Alt. °	Path Width km	Duration m:ss
21:29	180	00	−33	18.5	+56	177	3:39
21:52	−170	00	−37	23.0	+65	179	3:57
22:12	**−161**	**17.9**	**−39**	**57.6**	**+68**	**179**	**4:02 (GE)**
22:14	−160	00	−40	16.4	+68	178	4:02
22:34	−150	00	−41	56.2	+64	176	3:57
22:53	−140	00	−42	28.8	+57	173	3:43
23:09	−130	00	−42	01.8	+48	169	3:23
23:22	−120	00	−40	42.3	+38	163	3:02
23:33	−110	00	−38	38.1	+29	156	2:39
23:41	−100	00	−35	58.2	+19	147	2:17
23:45	−90	00	−32	53.3	+10	137	1:57
23:47	−80	00	−29	34.3	+01	127	1:42

Table 7.75 **Local Contact Times: 2012 November 13**

     a. Total

City/town	P1	U1	Maximum	U2	P2	Duration m:ss
Cairns, Australia	19:44	20:38	20:39	20:40	21:40	2:01

     b. Partial

City	P1	Maximum	P2	Magnitude
Adelaide, Australia	20:12	21:00	21:52	0.530
Auckland, New Zealand	20:18	21:27	22:43	0.870
Brisbane, Australia	19:56	20:54	21:58	0.835
Melbourne	20:16	21:06	22:00	0.525
Sydney	20:07	21:02	22:03	0.669
Wellington, New Zealand	20:26	21:33	22:47	0.764

Table 7.76 **Weather Prospects: 2012 November 14**

City	Clear	Scattered	Broken	Overcast
Cairns, Australia	0.6	48.8	45.1	5.6

ANNULAR SOLAR ECLIPSE OF 2013 MAY 9–10 (SAROS 138)

Australia and the South Pacific are once again the site for a central eclipse of the Sun (figure 7.26), but this time the Moon's umbra won't quite make it down to the surface. Instead, observers will see an annular eclipse.

Annularity begins at sunrise near Lake Disappointment, in the Gibson Desert of Western Australia. While the lake may be disappointing, the view of a ring of fire rising above the distant horizon should be spectacular. Moving northeastward, the eclipse's course passes into the Northern Territory, crossing over the towns of Tennant Creek and Wollogorang before hopping across the Gulf of Carpentaria and onto Cape York Peninsula, like the November 2012 event. This time passing north of Cairns, the central track visits the town of Musgrave before leaving the continent for the Coral Sea.

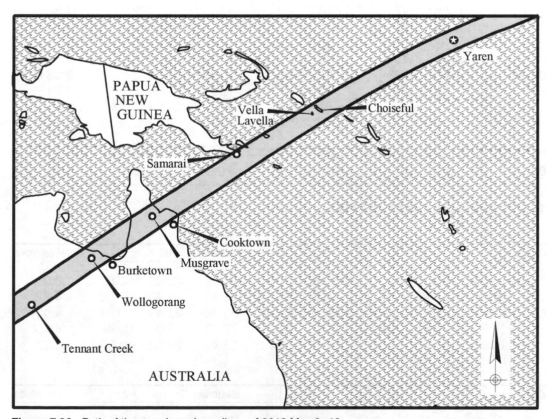

Figure 7.26   Path of the annular solar eclipse of 2013 May 9–10.

The next landfall occurs at the very tip of Papua New Guinea, where the seaside town of Samarai sees a central eclipse. Annularity then island-hops through the Solomon Islands, passing across Choiseul and Vella Lavella in the chain. Farther out at sea, Nauru also enjoys more than 5 minutes of annularity as the track engulfs the tiny island nation's residents.

The duration of central annularity will continue to increase until it reaches a maximum of 6 minutes 3 seconds at a point far from land in the South Pacific. The track will then begin to turn southeastward, avoiding any landfall for the rest of its journey.

Table 7.77 **Annular Solar Eclipse of 2013 May 9–10**
Center Line Coordinates

U.T. hh:mm	Longitude °	′	Latitude °	′	Alt. °	Path Width km	Duration m:ss
22:32	+120	00	−24	14.8	+01	222	4:11
22:34	+130	00	−20	39.1	+11	210	4:22
22:42	+140	00	−16	19.9	+23	198	4:37
22:57	+150	00	−11	13.7	+37	187	4:57
23:23	+160	00	−05	30.9	+54	176	5:25
00:01	+170	00	−00	04.8	+70	173	5:55
00:26	**+175**	**30.5**	**+02**	**12.3**	**+74**	**173**	**6:03 (GE)**
00:43	180	00	+03	30.8	+72	175	6:03
01:20	−170	00	+04	33.1	+57	181	5:45
01:47	−160	00	+03	35.3	+41	189	5:16
02:04	−150	00	+01	24.8	+27	200	4:51
02:13	−140	00	−01	24.3	+14	211	4:32
02:17	−130	00	−04	31.1	+03	223	4:18

Table 7.78 **Local Contact Times: 2013 May 9–10**
a. Annular

City/town	P1	U1	Maximum	U2	P2	Duration m:ss
Tennant Creek, Australia	21:25	22:35	22:37	22:38	00:02	3:08
Wollogorang, Australia	21:25	22:37	22:40	22:42	00:09	4:32
Musgrave, Australia	21:27	22:44	22:46	22:48	00:23	4:40
Samarai, Papua New Guinea	21:32	22:56	22:59	23:01	00:46	4:49
Yaren, Nauru	21:59	23:46	23:49	23:52	01:57	5:22

b. Partial

City	P1	Maximum	P2	Magnitude
Adelaide, Australia	21:39	22:44	23:59	0.502
Auckland, New Zealand	23:05	23:48	00:33	0.079

Table 7.78 **(Continued)**

City	P1	Maximum	P2	Magnitude
Brisbane, Australia	21:41	22:57	00:28	0.517
Hilo, Hawaii	00:27	01:52	03:05	0.473
Honolulu	00:22	01:47	03:00	0.440
Melbourne	21:50	22:52	00:02	0.365
Sydney	21:49	22:57	00:14	0.388

Table 7.79 **Weather Prospects: 2013 May 9–10**

City	Clear	Scattered	Broken	Overcast
Cooktown, Australia	Clouds < 3/10 = Not recorded		Precip > 0.1 inch = 16.5 days	
Yaren, Nauru	Clouds < 3/10 = Not recorded		Precip > 0.1 inch = 10.4 days	
Samarai, Papua New Guinea	Clouds < 3/10 = 10.9 days		Precip > 0.1 inch = 2.1 days	
Tennant Creek, Australia	31.0	44.6	21.6	2.7

ANNULAR/TOTAL SOLAR ECLIPSE OF 2013 NOVEMBER 3 (SAROS 143)

Five months and half a world later, the Moon's shadow again touches the Earth, this time giving us an annular-total eclipse that spans the Atlantic before hooking east-northeast toward Africa (figure 7.27). As with the annular-total eclipse of 2005 April 8, I feel obliged to post a warning up front that, as the eclipse date nears, your selection of observing site should be guided by the latest data published in one of the astronomical periodicals listed in Appendix B, or from a NASA Eclipse Bulletin (see Appendix G).

If only the world rotated just a little faster, the track would begin in the United States; but as it is, the shadow first touches down about 700 kilometers (440 miles) off the coast of North and South Carolina (so the East Coast will only see the Moon exiting the Sun's face at sunrise). Unlike the 2005 annular-total eclipse, where a good portion of the eclipse track produced an annular phase, this eclipse is a total event over all but the extreme ends of the center line. The central-eclipse path passes south of the Cape Verde Islands, then curves southeastward parallel to the African coastline. Greatest eclipse, with 1 minute 40 seconds of totality, occurs approximately 400 kilometers (250 miles) off the coast of Liberia.

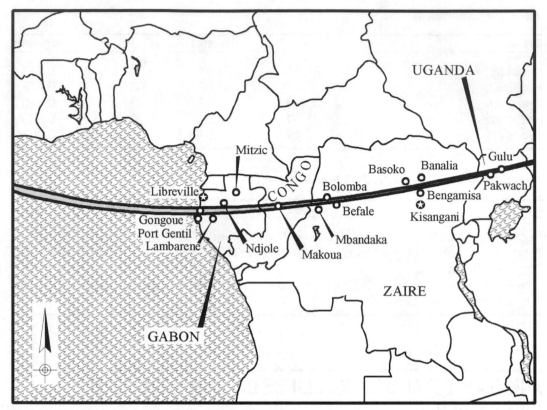

Figure 7.27 Track of the annular/total solar eclipse of 2013 November 3.

The shadow does not come ashore until it bumps into the nation of Gabon on the African west coast. The thin line of the central eclipse misses the coastal town of Port Gentil, instead passing over the nearby village of Gongoúe. Unfortunately, that's typical for this track; the path will avoid larger towns and cities, favoring instead small, isolated villages scattered within large, uninhabited stretches. This fact, combined with the unforgivingly narrow path, will make venturing to this eclipse an exercise in planning and forethought.

The central eclipse next passes into the northern portions of Congo, where it crosses over the town of Makoua. In Zaire, the path nips Befale, but just skims by nearby Bolomba. To their east, it misses the city of Kisangani (also known as Stanleyville) by about 100 kilometers (60 miles), but adventurous umbraphiles might venture on the main road leaving town toward the north. The path crosses this road about halfway between the villages of Bengamisa and Banalia, where totality will last only about half a minute.

In Uganda, the southern edge of the increasingly thin track breezes over Pakwach, Gulu, and Murchison Falls National Park, offering some interesting photographic possibilities. Finally the thin line of totality will briefly enter Kenya, crossing Lake Rudolph just north of Central Island, and Ethiopia. In the latter, the town of Mega will see a fleeting circle-of-light annular eclipse at sunset. The path leaves the Earth at the Ethiopia-Somalia border.

Table 7.80 **Annular/Total Solar Eclipse of 2013 November 3**
Center Line Coordinates

U.T. hh:mm	Longitude °	′	Latitude °	′	Alt. °	Path Width km	Duration m:ss
11:05	−70	00	+30	05.7	+01	4	0:04
11:07	−60	00	+26	56.4	+11	7	0:08
11:13	−50	00	+23	05.1	+22	18	0:23
11:25	−40	00	+18	26.9	+35	31	0:43
11:46	−30	00	+13	04.8	+49	43	1:06
12:16	−20	00	+07	28.7	+64	53	1:29
12:47	**−11**	**39.6**	**+03**	**29.4**	**+71**	**58**	**1:40 (GE)**
12:52	−10	00	+02	49.8	+71	57	1:39
13:26	+00	00	+00	11.4	+60	54	1:28
13:52	+10	00	−00	25.6	+46	45	1:06
14:10	+20	00	+00	25.8	+32	33	0:44
14:21	+30	00	+02	13.8	+19	21	0:25
14:26	+40	00	+04	36.0	+08	9	0:09

Table 7.81 **Local Contact Times: 2013 November 3**
a. Annular/Total

City/town	P1	U1	Maximum	U2	P2	Duration[1] m:ss
Gongoúe, Gabon	12:12	13:50	13:50	13:51	15:14	1:01
Makoua, Congo	12:31	14:03	14:03	14:03	15:20	0:43
Befale, Zaire	12:43	14:10	14:10	14:11	15:24	0:43
Bolomba, Zaire	12:40	—	14:09	—	15:23	0:00
Pakwach, Uganda	13:06	—	14:22	—	15:27	0:00
Gulu, Uganda	13:07	—	14:22	—	15:27	0:00
Mega, Ethiopia	13:15	—	14:26	—	15:27	0:00

Table 7.81 **(Continued)**
   b. Partial

City	P1	Maximum	P2	Magnitude
Addis Ababa, Ethiopia	13:14	14:24	—	0.861
Algiers, Algeria	12:14	12:55	13:35	0.101
Charleston, South Carolina	—	—	12:06	—
Johannesburg	13:39	14:15	14:50	0.134
Kisangani, Zaire	12:55	14:17	15:26	0.965
Lambarene, Gabon	12:15	13:52	15:15	0.998
Lisbon	11:35	12:22	13:10	0.145
Miami	—	—	12:02	—
Nairobi, Kenya	13:16	14:26	—	0.847
New York	—	—	12:10	—
Port Gentil, Gabon	12:11	13:49	15:13	0.999

1. Cities with durations of "0:00" are at the very edge of the total eclipse path, with totality passing almost instantaneously.

Table 7.82 **Weather Prospects: 2013 November 3**

City	Clear	Scattered	Broken	Overcast
Gulu, Uganda	Clouds < 3/10 = 0.8 day		Precip > 0.1 inch = 6.6 days	
Kisangani, Zaire	Clouds < 3/10 = 1.2 days		Precip > 0.1 inch = 10.0 days	
Libreville, Gabon	0	11.5	73.8	14.7
Mitzic, Gabon	Clouds < 3/10 = 0 days		Precip > 0.1 inch = 13.5 days	

ANNULAR SOLAR ECLIPSE OF 2014 APRIL 29 (SAROS 148)

April 2014 brings with it an unusual "point" annular eclipse. "Point" means that, because of the geometry involved, the eclipse will be visible from only a small area on the Earth's surface. As you can see in figure 7.28, that point happens to be within the uninhabited region of Wilkes Land in Antarctica! NASA's Fred Espenak explains the unusual circumstances that will create this eclipse:

The 2014 eclipse is one in which the central axis of the Moon's shadow misses the Earth, but the edge of the umbra hits Antarctica. This is called a "noncentral antumbral" eclipse in the technical jargon. And since noncen-

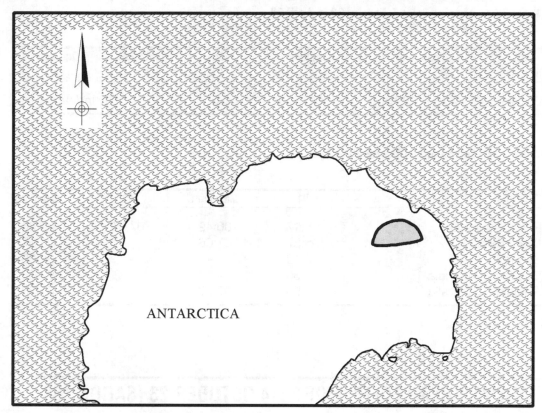

Figure 7.28 Path of the annular solar eclipse of 2014 April 29.

tral eclipses do not have a center line, you cannot quote a center-line duration. Just how long greatest eclipse will last will depend on how deeply into the shadow one goes (i.e., how close an observer gets to the shadow axis as it swings below the Earth). The longest duration will occur along the Earth's terminator, where the eclipse occurs at sunset. Atmospheric refraction, however, will play its largest role at this point of the eclipse, making accurate predictions impossible!

Restricted as annularity is, a partial eclipse will be visible over the greater part of Australia that day. The entire eclipse will only be visible west of a line southward from approximately the Cape York Peninsula to Spencer Gulf; to the east, the Sun will set before the end. Indeed, from the coastal cities of Brisbane, Sydney, and even Melbourne, the Sun will set at or before maximum eclipse.

Table 7.83 **Annular Solar Eclipse of 2014 April 29**
Center Line Coordinates

U.T. hh:mm	Longitude °	′	Latitude °	′	Alt. °	Path Width km	Duration m:ss
06:04	131	09.5	−70	41.8	0.5	—	Indeterminate

Table 7.84 **Local Contact Times: 2014 April 29**
Partial

City	P1	Maximum	P2	Magnitude
Perth, Australia	05:17	06:42	07:59	0.588
Melbourne	05:57	07:06	—	0.637
Sydney	06:13	—	—	0.523
Brisbane, Australia	06:30	—	—	0.362
Adelaide, Australia	05:56	07:07	—	0.603

PARTIAL SOLAR ECLIPSE OF 2014 OCTOBER 23 (SAROS 153)

The second eclipse of 2014 is a relatively shallow partial event, visible across much of North America and easternmost Siberia. Greatest eclipse, at magnitude 0.811, is reached at a point in M'Clintock Channel, off the east coast of Victoria Island, in Canada's Northwest Territories. The rest of us have to settle for a less spectacular show.

Table 7.85 **Local Contact Times: 2014 October 23**
(Greatest eclipse: Time 21:45 U.T., Latitude 71° 10.1'S, Longitude 97° 04.6'E, Mag. 0.811)

City	P1	Maximum	P2	Magnitude
Atlanta	21:59	—	—	0.388
Chicago	21:35	22:42	—	0.553
Denver	21:18	22:34	23:43	0.556
Edmonton	20:40	22:01	23:17	0.725
Fairbanks, Alaska	19:57	21:12	22:28	0.686
Juneau, Alaska	20:09	21:30	22:50	0.697

Table 7.85 **Local Contact Times: 2014 October 23**

City	P1	Maximum	P2	Magnitude
Los Angeles	21:07	22:27	23:39	0.452
Mexico City	22:31	23:08	23:43	0.121
New York	21:49	—	—	—
Nome, Alaska	19:46	20:56	22:09	0.598
San Francisco	20:51	22:15	23:31	0.504
Seattle	20:34	22:00	23:19	0.641
Vancouver	20:31	21:56	23:16	0.658

TOTAL SOLAR ECLIPSE OF 2015 MARCH 20 (SAROS 120)

This eclipse traces a primarily watery trail along the North Atlantic, beginning off the southern tip of Greenland and winding its way counterclockwise to the northeast, passing between Iceland and the United Kingdom (figure 7.29). First landfall is over the Faeroe Islands, where the path's southern boundary just embraces this Danish-owned group of eighteen islands. The islands' capital, Tórshavn, is the place to be for this particular event, where residents will see a little over 2 minutes of totality. Greatest eclipse, lasting 2 minutes 47 seconds, occurs not far to the islands' north, in the Norwegian Sea.

Crossing the Arctic Circle, totality does not strike dry land again until it pays a visit to sparsely inhabited Svalbard, a larger island group owned by Norway. All of the principal islands (Spitsbergen, Nordaustlandet, Barentsøya, Edgeøya, Kong Karls Land, Prins Karls Forland, and Bjørnøya) are within the broad shadow's track, although the best views will come from western locations. The town of Barentsburg, on Spitsbergen and along the center line, experiences totality for 2 minutes 30 seconds.

The Moon's shadow continues to hook counterclockwise toward the northwest. It finally leaves Earth very nearly at the North Pole, which, with the coming of the vernal equinox on the same date, will be seeing the Sun for the first time in six months. Quite an auspicious welcome! (Geometrically, the Sun is half above and half below the horizon from the North Pole at the vernal equinox. Our atmosphere bends, or refracts, the light to raise the Sun another 0.5°, bringing it completely up.)

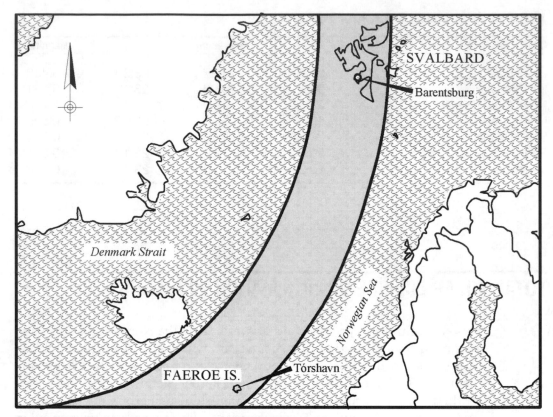

Figure 7.29 Track of the total solar eclipse of 2015 March 20.

Table 7.86 **Total Solar Eclipse of 2015 March 20**
Center Line Coordinates

U.T. hh:mm	Longitude °	′	Latitude °	′	Alt. °	Path Width km	Duration m:ss
09:13	−40	00	+53	46.0	+03	425	2:11
09:17	−30	00	+55	00.9	+10	466	2:26
09:26	−20	00	+57	48.2	+15	486	2:39
09:40	−10	00	+62	24.8	+18	472	2:46
09:46	**−06**	**34.1**	**+64**	**25.0**	**+19**	**462**	**2:47 (GE)**
09:55	00	00	+68	37.9	+18	441	2:44
10:07	+10	00	+75	27.8	+13	420	2:35
10:14	+20	00	+81	35.8	+08	412	2:23
10:17	+30	00	+85	46.2	+04	410	2:14
10:17	+40	00	+87	42.6	+02	409	2:11
10:18	+50	00	+88	30.9	+01	409	2:08
10:18	+60	00	+88	54.0	+01	409	2:06

Table 7.87   **Local Contact Times: 2015 March 20**

a. Total

City/town	P1	U1	Maximum	U2	P2	Duration m:ss
Tórshavn, Faeroe Islands	08:38	09:40	09:41	09:42	10:47	2:04
Barentsburg, Svalbard	09:10	10:09	10:10	10:11	11:11	2:30

b. Partial

City	P1	Maximum	P2	Magnitude
Berlin	08:38	09:47	10:58	0.789
London	08:24	09:30	10:40	0.868
Madrid	08:04	09:08	10:17	0.727
Moscow	09:12	10:19	11:26	0.653
Oslo	08:46	09:53	11:01	0.902
Paris	08:22	09:29	10:39	0.817
Rome	08:23	09:31	10:42	0.620

Table 7.88   **Weather Prospects: 2015 March 20**

City	Clear	Scattered	Broken	Overcast
Barentsburg, Svalbard	16.3	15.0	32.7	35.5
Tórshavn, Faeroe Islands	0.8	16.4	34.9	44.9

PARTIAL SOLAR ECLIPSE OF 2015 SEPTEMBER 13 (SAROS 125)

The second solar eclipse of 2015 is a minor partial event, visible only from portions of southern Africa and Antarctica. Greatest eclipse, where about three-quarters of the Sun will be covered, occurs in the Queen Maud Land region of Antarctica, while more temperate climes see a markedly less impressive event.

Table 7.89   **Local Contact Times: 2015 September 13**

(Greatest eclipse: Time 06:55 u.t., Latitude 72° 06.4′S, Longitude 02° 18.4′E, Mag. 0787)

City	P1	Maximum	P2	Magnitude
Cape Town	04:44	05:43	06:48	0.422
Johannesburg	04:42	05:34	06:32	0.263
Kerguelen Islands	05:57	07:14	08:32	0.384

TOTAL SOLAR ECLIPSE OF 2016 MARCH 9 (SAROS 130)

With this eclipse, the book has come full circle. Now, eighteen years after the Caribbean total eclipse of February 1998, comes the next eclipse in the same Saros. A quick comparison between these two eclipses' regional maps should immediately reveal the family resemblance. But, as described in chapter 1, the paths do not line up exactly; instead, this second track is off-set to the west of the earlier event.

This eclipse once again beckons eclipse chasers to Indonesia (figure 7.30). First touching down in the eastern Indian Ocean, the umbra quickly begins to island-hop. First the shadow crosses Sumatra, where a portion of the island's southern region experiences a brief morning total eclipse. Next, totality races across the smaller islands of Belitung and a portion of Bangka. Then it's on to southern Borneo, where the towns of Sampit, Balikpapan, and Amuntai each stand to see between 1 and 2 minutes of total-

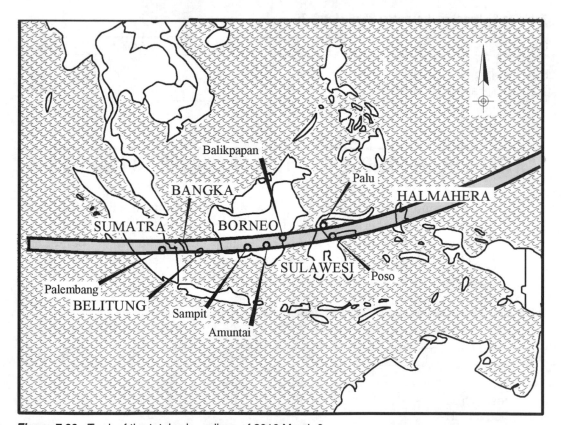

Figure 7.30   Track of the total solar eclipse of 2016 March 9.

ity. Farther along, the umbra lands on the island of Sulawesi, with the town of Poso and neighboring villages due to see nearly 3 minutes of totality. After a brief passage across the Molucca Sea, the umbra crosses Halmahera before it leaves Indonesia behind.

Passing into the South Pacific, the umbra crosses some of the tiny islands in Micronesia before heading out into the open ocean. The point of greatest eclipse, yielding 4 minutes 10 seconds of totality, occurs some 560 kilometers (350 miles) east of Manila.

The Moon's umbral shadow continues across the Pacific, finally leaving Earth at a watery point about 1,400 kilometers (900 miles) northeast of Hawaii. Hawaii itself sees a fairly deep partial eclipse as the Sun sets that evening. Alaska also sees a minor partial eclipse just before sunset.

Table 7.90 **Total Solar Eclipse of 2016 March 9**
Center Line Coordinates

U.T. hh:mm	Longitude °	′	Latitude °	′	Alt. °	Path Width km	Duration m:ss
00:16	+90	00	−02	23.0	+02	94	1:30
00:19	+100	00	−02	47.8	+12	107	1:54
00:25	+110	00	−02	31.3	+24	121	2:19
00:38	+120	00	−01	19.0	+37	134	2:50
00:58	+130	00	+01	09.8	+52	146	3:26
01:26	+140	00	+05	14.9	+67	154	3:58
01:58	**+148**	**50.2**	**+10**	**06.5**	**+75**	**155**	**4:10 (GE)**
02:01	+150	00	+10	48.2	+75	154	4:09
02:34	+160	00	+16	43.6	+64	149	3:52
02:59	+170	00	+21	50.1	+49	140	3:20
03:16	180	00	+25	46.5	+36	130	2:48
03:27	−170	00	+28	40.1	+24	120	2:22
03:34	−160	00	+30	42.7	+14	110	1:59
03:37	−150	00	+32	04.2	+05	101	1:40

Table 7.91 **Local Contact Times: 2016 March 9**
a. Total

City/town	P1	U1	Maximum	U2	P2	Duration m:ss
Palembang, Sumatra	23:20	00:20	00:21	00:22	01:31	1:59
Sampit, Borneo	23:22	00:27	00:28	00:29	01:44	2:11
Amuntai, Borneo	23:23	00:30	00:31	00:32	01:48	1:50
Balikpapan, Borneo	23:25	00:33	00:34	00:34	01:53	1:09
Poso, Sulawesi	23:27	00:38	00:39	00:40	02:01	2:43

Table 7.91 **(Continued)**
  b. Partial

City	P1	Maximum	P2	Magnitude
Cairns, Australia	00:12	01:12	02:15	0.288
Darwin, Australia	23:37	00:47	02:05	0.593
Hilo, Hawaii	02:37	03:37	—	0.638
Hong Kong	00:05	00:58	01:56	0.331
Honolulu	02:33	03:36	04:33	0.703
Manila	23:51	00:58	02:14	0.566
Singapore	23:22	00:23	01:32	0.886

Table 7.92 **Weather Prospects: 2016 March 9**

City	Clear	Scattered	Broken	Overcast
Balikpapan, Borneo	0.2	15.8	46.3	37.5
Palembang, Sumatra	0.4	12.6	66.5	19.6
Palu, Sulawesi	0	30.3	64.6	5.1

ANNULAR SOLAR ECLIPSE OF 2016 SEPTEMBER 1 (SAROS 135)

This second solar eclipse of 2016 will bring eclipse chasers back to southern Africa, as the region again hosts an annular event (figure 7.31). After first touching down in the Atlantic, about halfway between Africa and South America, the zone of annularity races eastward. It strikes land first in the village of Omboué, Gabon, where annularity lasts nearly 3 minutes. Though the track again avoids any sizable cities, it does cross the village of Mouila, on the main highway south of Gabon's capital, Libreville. The path continues across Gabon, passing over Franceville in the country's southeast corner, before heading into Congo, where it touches few, if any, accessible villages.

The path traverses the southern half of Lake Mai Ndombe before slowly turning southeastward into the jungles of central Zaire. The path again avoids cities in favor of small towns and villages. Dekese, on one of the country's main roads out of the capital, Kinshasa, rides the central line of annularity, but be forewarned that getting there is an awfully long drive!

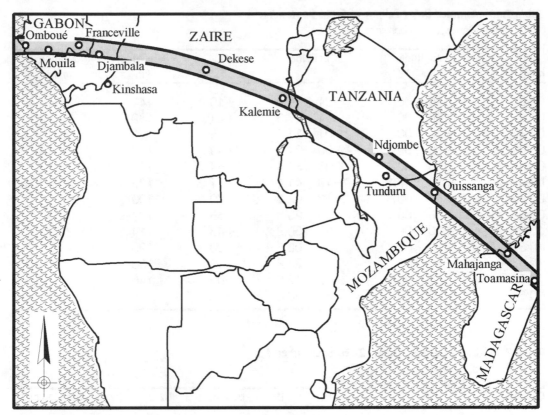

Figure 7.31   Path of the annular solar eclipse of 2016 September 1.

Kalemie, on the western shore of Lake Tanganyika, is a little south of the exact center line, but might be easier to get to, as it is one of the largest towns on the eclipse path.

Crossing the lake, the eclipse enters Tanzania as it heads toward the Indian Ocean. Traveling to the north of Lake Nyasa, the path crosses the villages of Njombe and Makumbako, near the Kipengere Mountains. Greatest eclipse, producing 3 minutes 6 seconds of annularity, occurs in southern Tanzania, northeast of the village of Tunduru (itself located south of the path of annularity).

Onward, the path pushes into northeastern Mozambique, where annularity visits Quissanga and other neighboring villages north of Pemba. Then a quick channel-hop brings the eclipse to Madagascar, the central path crossing diagonally from Mahajanga on the west coast to Toamasina on the east before plunging into the Indian Ocean.

Table 7.93 **Annular Solar Eclipse of 2016 September 1**
Center Line Coordinates

U.T. hh:mm	Longitude °	'	Latitude °	'	Alt. °	Path Width km	Duration m:ss
07:20	−10	00	−01	56.5	+10	143	2:44
07:27	00	00	−01	17.8	+21	132	2:48
07:40	+10	00	−01	31.7	+35	122	2:53
08:02	+20	00	−03	04.5	+49	113	2:59
08:35	+30	00	−06	31.4	+64	104	3:04
09:08	**+37**	**48.1**	**−10**	**41.4**	**+71**	**100**	**3:06 (GE)**
09:15	+40	00	−12	02.3	+70	100	3:06
09:53	+50	00	−18	20.3	+58	104	3:03
10:20	+60	00	−23	50.8	+43	114	2:59
10:37	+70	00	−28	08.8	+30	125	2:55
10:47	+80	00	−31	23.8	+19	136	2:52
10:52	+90	00	−33	48.6	+09	146	2:50
10:54	+100	00	−35	33.5	+01	158	2:47

Table 7.94 **Local Contact Times: 2016 September 1**
a. Annular

City/town	P1	U1	Maximum	U2	P2	Duration m:ss
Omboué, Gabon	06:18	07:38	07:39	07:40	09:17	2:50
Mouila, Gabon	06:20	07:41	07:42	07:43	09:22	2:35
Franceville, Gabon	06:22	07:45	07:46	07:48	09:29	2:25
Dekese, Zaire	06:32	08:05	08:06	08:08	09:58	3:00
Kalemie, Zaire	06:48	08:30	08:31	08:33	10:27	2:45
Njombe, Tanzania	07:05	08:53	08:54	08:55	10:49	2:01
Quissanga, Mozambique	07:24	09:16	09:18	09:19	11:08	3:04
Mahajanga, Madagascar	07:47	09:39	09:40	09:41	11:24	2:10
Toamasina, Madagascar	08:00	09:50	09:51	09:53	11:31	2:50

b. Partial

City	P1	Maximum	P2	Magnitude
Cape Town	07:42	08:48	09:58	0.261
Johannesburg	07:31	09:02	10:37	0.496
Kinshasa, Zaire	06:26	07:53	09:39	0.925
Nairobi, Kenya	07:00	08:46	10:38	0.760
Pretoria	07:30	09:02	10:38	0.508

Table 7.95 **Weather Prospects: 2016 September 1**

City	Clear	Scattered	Broken	Overcast
Franceville, Gabon	Clouds $< ^3/_{10}$ = 0.2 days		Precip > 0.1 inch = 7.2 days	
Kinshasa, Zaire	1.9	20.2	38.5	39.4
Mahajanga, Madgascar	Clouds $< ^3/_{10}$ = 22.9 days		Precip > 0.1 inch = 0.4 days	
Mayumba, Gabon	0.5	2.5	31.2	65.8
Mouila, Gabon	Clouds $< ^3/_{10}$ = 0 days		Precip > 0.1 inch = 3.3 days	
Njombe, Tanzania	Clouds $< ^3/_{10}$ = 2.6 days		Precip > 0.1 inch = 0.6 days	

ANNULAR SOLAR ECLIPSE OF 2017 FEBRUARY 26 (SAROS 140)

The fine thread of annularity again strikes south of Earth's Equator in this first solar eclipse of 2017. The central eclipse begins in the far southeast corner of the Pacific, off the coast of southern Chile (figure 7.32). Moving east-northeast, landfall occurs first along the southern shore of the Gulf of Corcovado. Moving inland, the track crosses the Chilean village of Puerto Aisén and the larger town of Coihaique before moving into Argentina. There, annularity briefly engulfs the towns of Malaspina and Camarones, both found along the coastal highway running from Comodoro Rivadavia and Rawson.

The path then moves out into the South Atlantic, bound for Africa. The path of annularity narrows from greater than 50 kilometers (31 miles) at Coihaique to a mere 31 kilometers (19 miles) at the greatest-eclipse point, which occurs about midway between the continents. There the Sun and Moon appear so close in size to one another that annularity lasts only 44 seconds, likely producing the broken annulus often associated with an annular-total event.

Hooking to the east, annularity invades the west coast of Africa at Lucira, Angola. The path's width having expanded to more than 70 kilometers (44 miles), annularity lasts just over 1 minute, with the Sun hanging low in the western sky. Moving inland, the path crosses the village of Cuima, south of Huambo, then rapidly moves into northwest Zambia before leaving Earth just west of Lubumbashi, Zaire, in a brilliant, ringlike sunset.

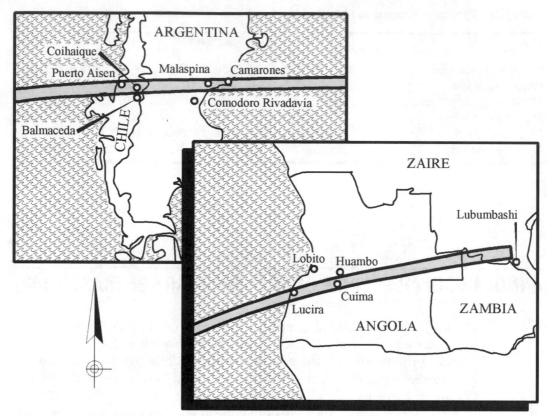

Figure 7.32 Path of the annular solar eclipse of 2017 February 26.

Table 7.96 **Annular Solar Eclipse of 2017 February 26**
Center Line Coordinates

U.T. hh:mm	Longitude °	'	Latitude °	'	Alt. °	Path Width km	Duration m:ss
13:16	−110	00	−43	41.6	+03	90	1:21
13:17	−100	00	−44	52.2	+10	83	1:18
13:22	−90	00	−45	35.5	+18	72	1:13
13:28	−80	00	−45	46.9	+26	63	1:09
13:38	−70	00	−45	20.5	+34	54	1:05
13:50	−60	00	−44	08.4	+43	45	0:59
14:07	−50	00	−42	00.0	+52	38	0:53
14:29	−40	00	−38	42.1	+59	33	0:47
14:54	**−31**	**08.4**	**−34**	**41.6**	**+63**	**31**	**0:44 (GE)**
14:56	−30	00	−34	06.2	+62	31	0:44

Table 7.96 **(Continued)**

U.T. hh:mm	Longitude °	 ′	Latitude °	 ′	Alt. °	Path Width km	Duration m:ss
15:26	−20	00	−28	30.6	+57	34	0:46
15:53	−10	00	−22	54.4	+44	44	0:52
16:12	00	00	−18	13.0	+31	57	1:00
16:24	+10	00	−14	41.6	+19	70	1:08
16:29	+20	00	−12	12.1	+07	82	1:14

Table 7.97 **Local Contact Times: 2017 February 26**

a. Annular

City/town	P1	U1	Maximum	U2	P2	Duration m:ss
Puerto Aisén, Chile	12:22	13:34	13:35	13:35	14:55	0:57
Coihaique, Chile	12:23	13:35	13:36	13:36	14:56	1:02
Camarones, Argentina	12:26	13:42	13:43	13:43	15:06	0:51
Malaspina, Argentina	12:26	13:41	13:41	13:42	15:04	0:53
Lucira, Angola	15:15	16:25	16:26	16:26	17:28	1:05
Cuima, Angola	15:19	16:27	16:27	16:28	17:28	1:10

b. Partial

City	P1	Maximum	P2	Magnitude
Asunción, Paraguay	12:41	13:57	15:21	0.453
Buenos Aires	12:31	13:52	15:22	0.720
Cape Town	14:58	15:59	16:54	0.518
Johannesburg	15:13	16:13	—	0.622
Lubumbashi, Angola	15:28	—	—	0.969
Pretoria	15:14	16:13	—	0.629
Rio de Janeiro	13:08	14:39	16:10	0.563

Table 7.98 **Weather Prospects: 2017 February 26**

City	Clear	Scattered	Broken	Overcast
Balmaceda, Chile	9.4	44.1	38.6	7.9
Comodoro Rivadavia, Argentina	Clouds $< ^3/_{10}$ = 10.2 days		Precip > 0.1 inch = 0.6 days	
Lobito, Angola	Clouds $< ^3/_{10}$ = 9.4 days		Precip > 0.1 inch = 3.4 days	
Puerto Aisén, Chile	Clouds $< ^3/_{10}$ = 2.9 days		Precip > 0.1 inch = 12.1 days	

TOTAL SOLAR ECLIPSE OF 2017 AUGUST 21 (SAROS 145)

If you live in the United States or Canada and you're like me, you are already anxiously awaiting this eclipse. Although the twentieth century featured no fewer than sixteen total solar eclipses crisscrossing portions of North America, the Moon's umbra has not touched the United States since February 1979. Nearly four decades later, the greatest celestial show on Earth will finally return. (It's interesting to note that, though we may think it rare that a total solar eclipse is visible from the United States, Americans of the twenty-third century will see total eclipses cross the country in 2245, 2252, 2254, 2259, 2261, and 2263, as well as annular eclipses in 2247 and 2251!)

The shadow first strikes the Earth not on land, but in the North Pacific Ocean, about halfway between the Aleutians and Hawaii. It quickly heads for the contiguous United States, where it cuts diagonally from northwest to southeast.

First to see totality is the area around Lincoln City, Oregon, with totality there lasting a little less than two minutes (figure 7.33a). Farther inland, totality passes almost centrally over Salem, Oregon, the first of many large cities to see the shadow of the Moon that day. After scaling the slopes of the Cascade Mountains, the shadow hitches a ride along U.S. Highway 26, bound for Idaho. Though it bypasses the capital of Boise, the total eclipse is seen from Idaho Falls, which lies on the southern boundary of totality, producing a relatively short total phase; observers there might do well to travel northeast toward Rexburg to see a longer duration.

Next on the eclipse's itinerary is Wyoming, where it strikes Jackson, then rejoins eastbound U.S. Highway 26 (figure 7.33b). The center line passes directly over the city of Casper, bringing with it almost two and a half minutes of totality before it races off toward Nebraska. Just across the state line, Scottsbluff, Nebraska, encounters the umbra for 1 minute 43 seconds before it begins its trek across the great plains for Lincoln. There, observers see a relatively brief 1-minute-22-second eclipse, the center line passing to the city's south. A better view will be had by heading south on U.S. Highway 77 toward Beatrice, where totality will last more than a minute longer.

The shadow then pushes into Missouri, passing centrally over both St. Joseph and Jefferson City. Kansas City will also be touched by the umbra, albeit ever so briefly, as the city lies on the southern edge of totality. St. Louis, on the shadow's northern edge, will also have a brief encounter with totality.

The southern tip of Illinois is next to host totality; the umbra covers an area from approximately Mount Vernon to Mounds (figure 7.33c). Ap-

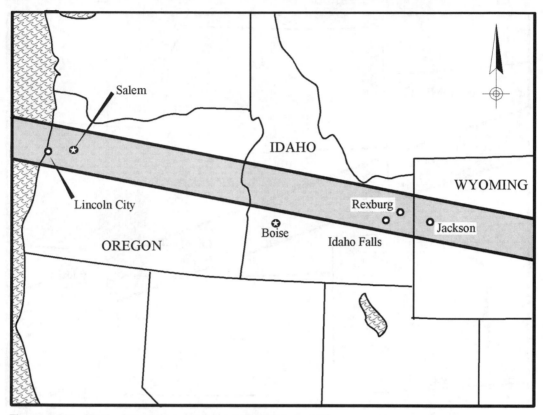

Figure 7.33a Track of the total solar eclipse of 2017 August 21.

proaching the point of greatest eclipse, Illinoisans in and around Carbondale and Marion witness more than 2 and a half minutes of totality.

Greatest eclipse, in fact, occurs near the small town of Crofton in western Kentucky, bringing with it 2 minutes 40 seconds of totality. Not far to the east, Bowling Green finds itself set on the northern edge of the total-eclipse zone, providing viewers a nearly instantaneous glimpse at the Sun's corona and photosphere.

The umbra passes through Tennessee from the city of Clarksville, through Nashville, and on to the south of Knoxville. The total eclipse then crosses into extreme western North Carolina, passing over the Great Smoky Mountains National Park.

In early afternoon the eclipse reaches its final state, South Carolina, before it marches to the sea. Spartanburg just makes it into the eclipse zone for a momentary encounter with totality, while both Greenville and Columbia lie along the center line, affording their citizens some of the best seats in the house. The umbra finally departs the continent at Charleston. Con-

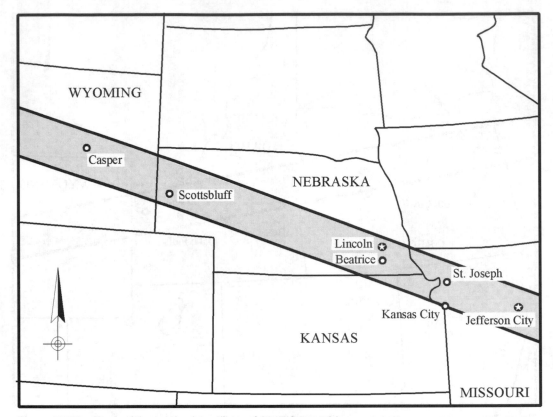

Figure 7.33b   Track of the total solar eclipse of 2017 August 21.

tinuing into the Atlantic, the eclipse does not quite make it to Africa before
leaving the Earth at a point south of the Cape Verde Islands.

Table 7.99   **Total Solar Eclipse of 2017 August 21**
Center Line Coordinates

U.T. hh:mm	Longitude °	'	Latitude °	'	Alt. °	Path Width km	Duration m:ss
16:48	−170	00	+40	03.1	+01	62	0:51
16:50	−160	00	+41	58.6	+09	72	1:05
16:54	−150	00	+43	29.5	+17	80	1:18
17:00	−140	00	+44	30.2	+26	88	1:33
17:09	−130	00	+44	55.0	+34	95	1:49
17:21	−120	00	+44	37.4	+42	102	2:05
17:37	−110	00	+43	29.1	+50	107	2:22

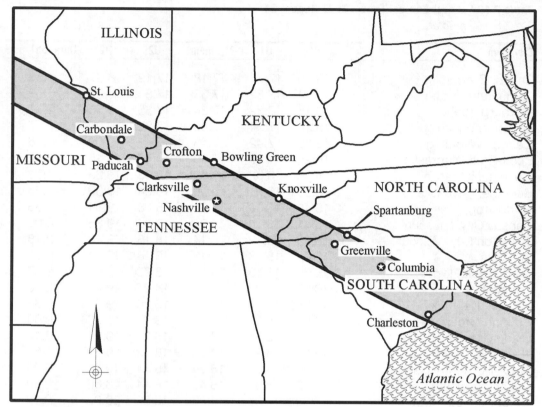

Figure 7.33c Track of the total solar eclipse of 2017 August 21.

Table 7.99 **(Continued)**

U.T. hh:mm	Longitude °	′	Latitude °	′	Alt. °	Path Width km	Duration m:ss
17:56	−100	00	+41	19.4	+59	111	2:34
18:19	−90	00	+37	56.8	+64	114	2:40
18:26	**−87**	**37.7**	**+36**	**58.0**	**+64**	**115**	**2:40 (GE)**
18:45	−80	00	+33	18.7	+61	114	2:35
19:12	−70	00	+27	51.1	+52	111	2:16
19:35	−60	00	+22	28.4	+39	102	1:51
19:50	−50	00	+17	53.3	+25	88	1:26
19:58	−40	00	+14	17.2	+13	74	1:07
20:01	−30	00	+11	35.0	+03	57	0:47

Table 7.100 **Local Contact Times: 2017 August 21**

a. Total

City/town	P1	U1	Maximum	U2	P2	Duration[1] m:ss
Salem, Oregon	16:05	17:17	17:18	17:19	18:37	1:53
Idaho Falls, Idaho	16:14	17:32	17:33	17:34	18:57	1:50
Rexburg, Idaho	16:15	17:32	17:34	17:35	18:58	2:17
Jackson, Wyoming	16:16	17:34	17:36	17:37	19:00	2:17
Casper, Wyoming	16:22	17:42	17:43	17:44	19:09	2:26
Scottsbluff, Nebraska	16:25	17:47	17:48	17:49	19:15	1:42
Lincoln, Nebraska	16:36	18:02	18:02	18:03	19:29	1:22
Beatrice, Nebraska	16:36	18:01	18:03	18:04	19:30	2:35
St. Joseph, Missouri	16:40	18:06	18:07	18:08	19:34	2:38
Kansas City, Missouri	16:41	18:08	18:08	18:08	19:35	0:00
Jefferson City, Missouri	16:45	18:12	18:14	18:15	19:40	2:18
St. Louis	16:49	18:18	18:18	18:18	19:44	0:00
Carbondale, Illinois	16:52	18:19	18:21	18:22	19:47	2:37
Paducah, Kentucky	16:53	18:22	18:23	18:24	19:49	2:20
Crofton, Kentucky	16:56	18:24	18:25	18:26	19:51	2:34
Bowling Green, Kentucky	16:58	18:27	18:27	18:27	19:52	0:00
Clarksville, Tennessee	16:56	18:25	18:26	18:27	19:52	2:17
Nashville, Tennessee	16:58	18:27	18:28	18:29	19:53	1:51
Knoxville, Tennessee	17:04	18:34	18:34	18:34	19:58	0:00
Columbia, South Carolina	17:12	18:41	18:42	18:44	20:06	2:30
Greenville, South Carolina	17:08	18:37	18:38	18:39	20:02	2:13
Spartanburg, South Carolina	17:09	18:39	18:39	18:39	20:03	0:00
Charleston, South Carolina	17:16	18:46	18:46	18:47	20:09	1:28

b. Partial

City	P1	Maximum	P2	Magnitude
Atlanta	17:05	18:36	20:01	0.971
Boston	17:28	18:46	19:59	0.702
Chicago	16:54	18:19	19:42	0.889
Dallas	16:40	18:09	19:39	0.801
Denver	16:23	17:46	19:14	0.934
Detroit	17:03	18:27	19:47	0.831
El Paso, Texas	16:23	17:46	19:15	0.683
Los Angeles	16:05	17:20	18:44	0.694
Miami	17:26	18:58	20:20	0.822
Montreal	17:21	18:38	19:50	0.662
New Orleans	16:57	18:29	19:57	0.799
New York	17:22	18:44	20:00	0.769
Philadelphia	17:20	18:44	20:01	0.800

1. Cities with durations of "0:00" are at the very edge of the total eclipse path, with totality passing almost instantaneously.

Table 7.100 **(Continued)**

City	P1	Maximum	P2	Magnitude
Phoenix	16:13	17:33	19:00	0.700
Salt Lake City	16:13	17:33	18:59	0.925
San Francisco	16:01	17:15	18:36	0.802
Seattle	16:08	17:20	18:38	0.930
Toronto	17:10	18:31	19:49	0.761
Vancouver	16:09	17:20	18:37	0.884
Washington, D.C.	17:17	18:42	20:01	0.845

Table 7.101 **Weather Prospects: 2017 August 21**

City	Clear	Scattered	Broken	Overcast
Casper, Wyoming	37.6	29.8	17.6	14.9
Charleston, South Carolina	2.5	28.6	43.4	25.5
Greenville, South Carolina	8.3	37.1	36.1	18.5
Lincoln, Nebraska	23.3	28.3	24.9	23.5
Nashville, Tennessee	7.9	40.1	32.6	19.5
Paducah, Kentucky	9.7	43.9	27.8	18.7
Salem, Oregon	33.4	14.0	16.9	35.7
Scottsbluff, Nebraska	35.6	29.0	19.9	15.6
St. Louis	11.7	33.6	31.1	23.6

There you have it—a quick look at the next twenty years' worth of solar eclipses. Many of the eclipses will pass through the most remote areas on Earth, while others cross some of the most densely populated regions. But regardless of where an eclipse will occur, someone will probably be there to enjoy it.

8 Lunar Eclipses: 1998–2017

Just as there are many exciting solar eclipses in the next two decades, so too are there some wonderful lunar eclipses. Unlike a total solar eclipse, where location is everything, the breathtaking impact of a total lunar eclipse can be enjoyed by millions of people across the night side of the Earth.

Following a format similar to the last chapter's treatment of solar eclipses, the lunar eclipses listed here are presented in chronological order. Each listing features a general discussion of the event's visibility followed by the Universal Times (U.T.; see Appendix E) of penumbral and, if applicable, umbral contacts. In the following timetables, the point when the Moon first enters Earth's penumbra is abbreviated P1, and P2 denotes its exit. Likewise, U1 and U4 specify when the Moon enters and leaves the darker umbral shadow. U2 and U3 denote the beginning and end of totality, respectively. (As mentioned in chapter 1, partial eclipses do not have U2 and U3 values, because those are reserved for when the Moon fully enters the Earth's umbra. Rather, contact times for partial lunar eclipses are designated only with P1, U1, U4, and P2, for first penumbral contact, first umbral contact, last umbral contact, and last penumbral contact. Likewise, penumbral eclipses have only P1 and P2 contact times.)

An eclipse's magnitude, or penetration into Earth's shadow, is also listed in each timetable. If the event is a total or partial umbral eclipse, the number refers to the Moon's encroachment into the umbra, and is designated *UMag*. The magnitudes of penumbral eclipses are denoted in the tables by *PMag*.

Each description of a total umbral eclipse is accompanied by a diagram depicting the Moon's track through Earth's shadow, and by a world map showing the areas where the eclipse will be visible. Note that the world maps are centered on the intersection of the Earth's prime meridian and Equator. Regions shaded in dark gray will see the entire eclipse from start to finish. Only part of the eclipse will be seen in the light gray regions that border the central zone, as the Moon will either rise or set sometime during the event. A general rule: areas to the left (or west) of the "entire-eclipse zone" will see the Moon rise already in eclipse, but regions to the right (typically east) will see the Moon set still shadowed by the Earth. (These directions may change depending on where the eclipsed region lies relative to the maps' edges.) Unshaded areas will not see the eclipse.

With all of this in mind, let's begin the review of things to come.

PENUMBRAL LUNAR ECLIPSE OF 1998 MARCH 13

This chapter begins not with a bang, but a whimper: a penumbral lunar eclipse. The first lunar eclipse of 1998 occurs with the Moon in the constellation Leo, nestled about halfway between the bright stars Regulus and Spica. At maximum eclipse, the penumbral magnitude will reach 0.74, when the southern three-quarters of the lunar disk will be immersed in the light gray penumbral portion of the Earth's shadow. The event will be visible to observers in North and South America, although only a slight shading of the Moon's southern limb will likely be noticed. The best chance of seeing any decrease in the Moon's brightness will be between 04:00 and 04:40 U.T.

Table 8.1   **Contact Times: 1998 March 13**

Type	P1	U1	U2	Max	U3	U4	P2	PMag
Penumbral	02:15	—	—	04:20	—	—	06:26	0.74

PENUMBRAL LUNAR ECLIPSE OF 1998 AUGUST 8

This is what you might call a "technical" lunar eclipse. Tonight, while in the constellation Capricornus, the Moon will *technically* pass through the

Earth's penumbra, but it is doubtful that any effect will be noticeable. That is because, even at maximum eclipse, the penumbral magnitude will reach only 0.15, far too little for any shading to be seen. This non-event will occur with the Moon in the sky over western Europe, Africa, South America, and eastern North America.

Table 8.2 **Contact Times: 1998 August 8**

Type	P1	U1	U2	Max	U3	U4	P2	PMag
Penumbral	01:32	—	—	02:25	—	—	03:18	0.15

PENUMBRAL LUNAR ECLIPSE OF 1998 SEPTEMBER 6

The third and final lunar eclipse of 1998 is another penumbral event, staged directly south of the Pleiades star cluster in Taurus the Bull. At maximum eclipse, 11:11 u.t., a little more than 80 percent of the Moon's disk will be immersed in the Earth's penumbra. This will create just a hint of shading on the northern lunar limb that should be visible to observers around the Pacific Rim.

Table 8.3 **Contact Times: 1998 September 6**

Type	P1	U1	U2	Max	U3	U4	P2	PMag
Penumbral	09:15	—	—	11:11	—	—	13:07	0.84

PENUMBRAL LUNAR ECLIPSE OF 1999 JANUARY 31

The year 1999 begins with a deep penumbral lunar eclipse in the constellation Cancer. Though penumbral eclipses usually pass all but unnoticed, this event should produce an obvious gray shading along the Moon's northern edge between about 15:30 and 16:50 u.t, as the lunar disk just barely misses the Earth's umbra. At that time the Moon will be highest in the sky for observers in Asia and Australia.

Table 8.4 **Contact Times: 1999 January 31**

Type	P1	U1	U2	Max	U3	U4	P2	PMag
Penumbral	14:05	—	—	16:18	—	—	18:31	1.03

PARTIAL LUNAR ECLIPSE OF 1999 JULY 28

Observers situated around the Pacific rim are favored to see the Moon slip partway into the Earth's umbra for the first time since 1997. At maximum, most of the Moon's southern half will be immersed in the umbra. The eclipse will occur in the constellation Capricornus, just west of the tiny triangle made up of the stars Rho, Pi, and Omicron Capricorni.

This will also be a good opportunity, especially for those who have never seen it, to spot the distant planet Neptune. During the eclipse, Neptune will be less than 2° west of the Moon, just east of the 5.5-magnitude star Sigma (σ) Capricorni. Although light from the Moon will be great, even at mid-eclipse, most telescopes should be able to reveal Neptune's pale turquoise eighth-magnitude disk.

Table 8.5 **Contact Times: 1999 July 28**

Type	P1	U1	U2	Max	U3	U4	P2	UMag
Partial	08:57	10:22	—	11:34	—	12:46	14:12	0.40

TOTAL LUNAR ECLIPSE OF 2000 JANUARY 21

The first total lunar eclipse in nearly three years will be seen tonight from most of the western hemisphere. The entire event will be visible from western Europe, western Africa, and all of North and South America (figure 8.1). Only extreme western Alaska, where the Moon will rise already in eclipse, will miss part of the show. Observers in central and eastern Europe and Asia will watch the Moon set before the eclipse ends, while the eclipse will occur during the day for southeast Asia and Australia.

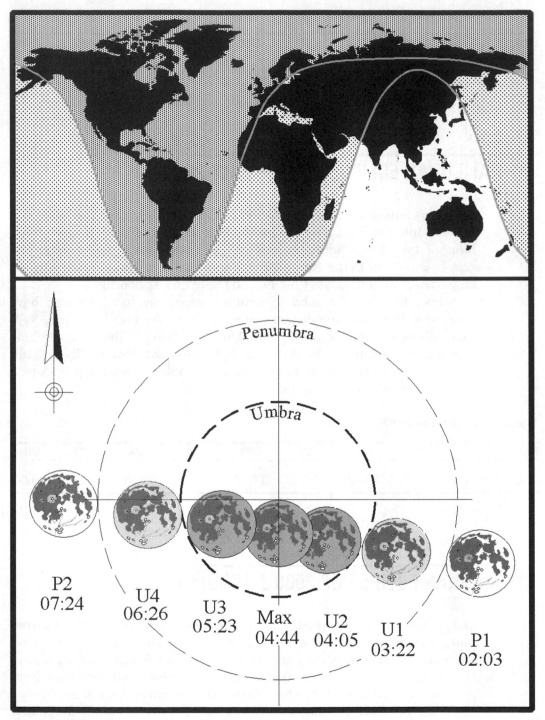

Figure 8.1 Circumstances of the total lunar eclipse of 2000 January 21.

The eclipse will occur with the Moon in the constellation of Cancer. Although no bright planets will be nearby, the Moon will lie about 8° west of the Beehive open star cluster (M44), and some 10° southeast of the twin stars Castor and Pollux in Gemini. At mid-eclipse, the umbral magnitude will peak at 1.33. This is a moderately deep value that could well result in a darker-than-average appearance to the lunar disk.

Table 8.6 **Contact Times: 2000 January 21**

Type	P1	U1	U2	Max	U3	U4	P2	UMag
Total	02:03	03:22	04:05	04:44	05:23	06:26	07:24	1.33

TOTAL LUNAR ECLIPSE OF 2000 JULY 16

The family of twentieth-century lunar eclipses will close with a total event best seen from eastern Asia, Australia, and New Zealand. As figure 8.2 shows, observers in the western halves of North and South America will also see the eclipse, although the Moon will set still immersed in Earth's shadow. Observers along the Pacific coast will enjoy the best vantage points. Central Asia and eastern Africa will see the Moon rise covered by at least part of the Earth's shadow, while Europe and the eastern halves of North and South America will miss the show entirely.

During the eclipse, the Moon will lie in eastern Sagittarius, about 12° east of the Teapot asterism and not far from the border with Capricornus. At mid-eclipse, the Moon will be near the center of the Earth's shadow, producing an umbral magnitude of 1.77. This is the deepest value of any eclipse in this book, and may result in a dark total phase. If that happens, long-exposure photographs made with wide-angle lenses may capture a pretty scene, with the fully eclipsed Moon on the eastern side of the film's frame, and some of the Sagittarius star clouds on the western edge.

Table 8.7 **Contact Times: 2000 July 16**

Type	P1	U1	U2	Max	U3	U4	P2	UMag
Total	10:47	11:58	13:02	13:56	14:49	15:54	17:05	1.77

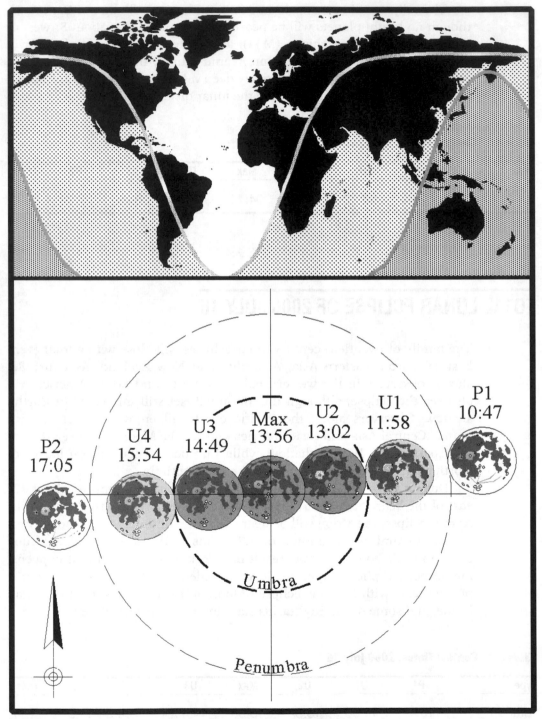

Figure 8.2 Circumstances of the total lunar eclipse of 2000 July 16.

TOTAL LUNAR ECLIPSE OF 2001 JANUARY 9

Kicking off the twenty-first century for lunar-eclipse observers in Europe, Asia, and much of Africa will be this total event. Portions of the eclipse will also be visible from eastern North and South America, as well as from Australia. As shown in figure 8.3, the lunar disk will hug the umbra's northern edge as it passes through Earth's shadow. Barring any outside interference, such as from an unforeseen volcanic eruption here on Earth, this will likely produce a relatively bright total phase.

Photographers with wide-angle lenses should find it an attractive scene to capture on film. The Moon will be within the constellation of Gemini during the eclipse, nestled between the familiar stick figures of the brothers. The planets Jupiter and Saturn will also be in the sky, some 45° to the Moon's west in Taurus the Bull.

Table 8.8 **Contact Times: 2001 January 9**

Type	P1	U1	U2	Max	U3	U4	P2	UMag
Total	17:44	18:43	19:50	20:21	20:52	22:00	22:58	1.20

PARTIAL LUNAR ECLIPSE OF 2001 JULY 5

The eastern hemisphere will be favored for this partial eclipse, where the northern half of the lunar disk will be bathed in the Earth's umbra. Australia, New Zealand, and the islands of Malaysia and Indonesia, as well as Japan and the eastern portions of mainland Asia, will have the best view, while central and western Asia, and even easternmost Africa, will see a portion of the eclipse.

The show's celestial stage will be set adjacent to the Teapot asterism in Sagittarius. To the Moon's west will be the red planet, Mars, beckoning observers to break away from the eclipse for a quick look.

Table 8.9 **Contact Times: 2001 July 5**

Type	P1	U1	U2	Max	U3	U4	P2	PMag
Partial	12:11	13:36	—	14:56	—	16:16	17:40	0.50

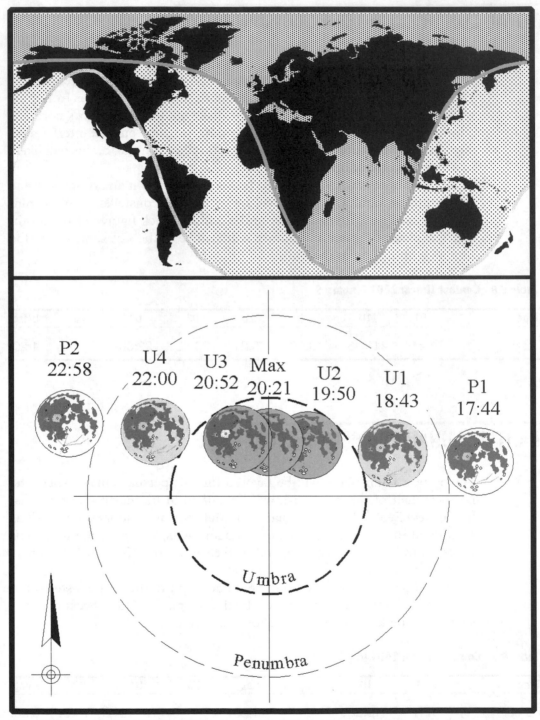

Figure 8.3 Circumstances of the total lunar eclipse of 2001 January 9.

PENUMBRAL LUNAR ECLIPSE OF 2001 DECEMBER 30

A penumbral eclipse visible from North America as well as from northeastern Asia will close out the year 2001. At maximum, all but the northern edge of the Moon will be immersed in the Earth's penumbra, producing a subtle darkening along the southern lunar limb. Look between approximately 10:10 and 10:50 U.T.

Table 8.10 **Contact Times: 2001 December 30**

Type	P1	U1	U2	Max	U3	U4	P2	PMag
Penumbral	08:26	—	—	10:30	—	—	12:34	0.92

PENUMBRAL LUNAR ECLIPSE OF 2002 MAY 26

This penumbral eclipse of the Moon will be visible over nearly all of the Pacific Ocean, including Hawaii, Australia, and New Zealand, although it should attract little attention except from die-hard eclipse observers. Even at maximum, the eclipse's penumbral magnitude will reach only 0.71, with the southern lunar limb dipping deepest into the tenuous shadow. This will likely produce a negligible shading visible over the Moon's south polar region from between about 11:50 and 12:20 U.T. Maximum shadow coverage occurs at 12:04 U.T.

Table 8.11 **Contact Times: 2002 May 26**

Type	P1	U1	U2	Max	U3	U4	P2	PMag
Penumbral	10:14	—	—	12:04	—	—	13:54	0.71

PENUMBRAL LUNAR ECLIPSE OF 2002 JUNE 24

One lunation later, another penumbral lunar eclipse will occur. While this event will be best seen from much of Europe, Asia, and Africa, its circumstances will be even less favorable than the preceding eclipse. Maximum eclipse, which will occur at 21:28 U.T., will reach a penumbral magnitude of only 0.24. With such a slim portion of the northern limb cutting into the penumbra, it is unlikely that any dimming of the Moon's brilliance will be discernible.

Table 8.12 **Contact Times: 2002 June 24**

Type	P1	U1	U2	Max	U3	U4	P2	PMag
Penumbral	20:19	—	—	21:28	—	—	22:36	0.24

PENUMBRAL LUNAR ECLIPSE OF 2002 NOVEMBER 20

A rather disappointing year, lunar-eclipse-wise, will close with a third penumbral event visible from both sides of the Atlantic Ocean. The eclipse will reach its greatest penumbral magnitude, 0.89, at 01:47 U.T. For perhaps fifteen or twenty minutes on either side of that time, a slight darkening along the Moon's northern limb might be detectable from Europe, western Africa, and eastern North and South America, especially through binoculars and low-power telescopes.

Table 8.13 **Contact Times: 2002 Nov 20**

Type	P1	U1	U2	Max	U3	U4	P2	PMag
Penumbral	23:32	—	—	01:47	—	—	04:01	0.89

TOTAL LUNAR ECLIPSE OF 2003 MAY 16

The year 2003 will feature a pair of total lunar eclipses, with this first event visible in its entirety from the eastern third of the United States and Canada as well as from all of South America. The Moon will rise already partially eclipsed for observers west of a line drawn from the southeastern tip of Hudson Bay to New Orleans, while amateurs in Europe and Africa will see the Moon set before it leaves Earth's shadow.

At maximum, umbral magnitude will reach 1.13. As shown in figure 8.4, the Moon will track to the north of the umbra's center. The eclipse will occur in Libra, some 15° northeast of the bright star Antares in Scorpius. Unfortunately, there will be no bright stars or planets in the immediate neighborhood to add to the visual interest of the event.

Table 8.14 **Contact Times: 2003 May 16**

Type	P1	U1	U2	Max	U3	U4	P2	UMag
Total	01:06	02:03	03:14	03:40	04:07	05:18	06:15	1.13

TOTAL LUNAR ECLIPSE OF 2003 NOVEMBER 8–9

Six months later, another total lunar eclipse will occur for Europe, northwestern Africa, and eastern North and South America (figure 8.5). This time the Moon will be situated in southeastern Aries, an area of sky not noted for bright stars or, at the time of the eclipse, planets. Still, a colorful event may greet observers.

The Moon will only slip through the southern portion of the umbra, producing a short total phase lasting less than half an hour. At maximum, which will occur at 01:19 u.t., the eclipse's umbral magnitude will be 1.02. Unless airborne volcanic aerosols or other atmospheric effects influence its appearance, the lunar disk should remain relatively bright, especially along its southern rim.

Table 8.15 **Contact Times: 2003 November 8–9**

Type	P1	U1	U2	Max	U3	U4	P2	UMag
Total	22:15	23:33	01:07	01:19	01:31	03:05	04:22	1.02

TOTAL LUNAR ECLIPSE OF 2004 MAY 4

The year 2004 will bring with it a pair of "twin" total lunar eclipses, each reaching an umbral magnitude of 1.31 at maximum. As shown in figure

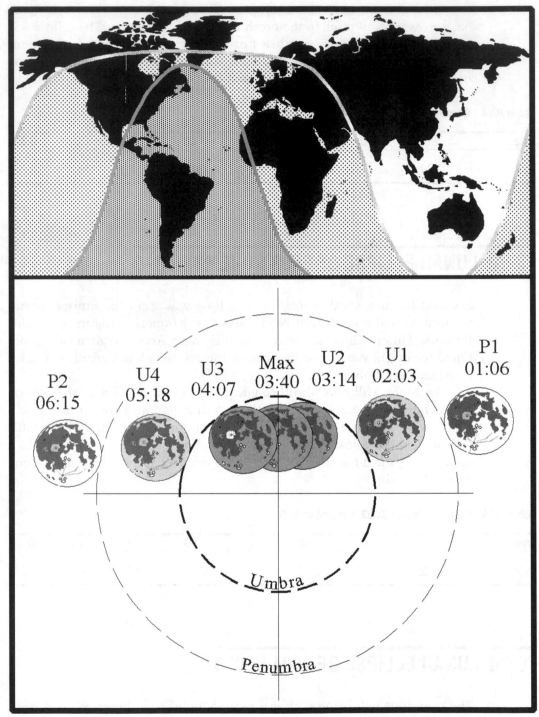

Figure 8.4 Circumstances of the total lunar eclipse of 2003 May 16.

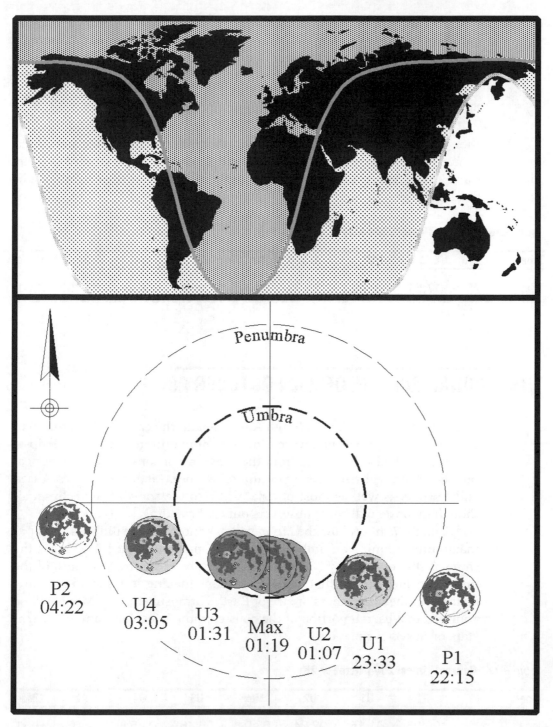

Figure 8.5 Circumstances of the total lunar eclipse of 2003 November 8–9.

8.6, sky watchers in central Asia and most of Africa will be able to monitor the progress of this first event from start to finish. Only the early stages of the umbral passage will be seen from eastern Asia, Australia, and New Zealand before the Moon sets at sunrise, while most of Europe and western Africa will watch the Moon rise that evening already partially eclipsed. North America will miss out on the show.

During the eclipse, the Moon will lie within the constellation Libra. Although no planets will be in the vicinity, the bright, wide double star Alpha Librae will be seen to the Moon's north during totality. Observers near Cape Town, South Africa, will see a spectacular show as the Moon's northern limb grazes the star.

Table 8.16 **Contact Times: 2004 May 4**

Type	P1	U1	U2	Max	U3	U4	P2	UMag
Total	17:51	18:49	19:53	20:31	21:09	22:13	23:10	1.31

TOTAL LUNAR ECLIPSE OF 2004 OCTOBER 28

Western Europe and Africa, South America, and the eastern three-quarters of North America will be favored for this mid-autumn total lunar eclipse. The event will also be visible from the western portions of North America, but the Moon will rise already in umbral eclipse. Likewise, eastern Africa and Europe, as well as most of Asia, will also see some of the eclipse, although moonset will cheat observers out of the ending.

Figure 8.7 shows how the Moon will pass north of the umbra's center at maximum, attaining an umbral-penetration magnitude of 1.31. This is the same as that of the May 2004 event (though the Moon travels south of the umbra's center during that eclipse), perhaps making it easier to spot any changes in the opaqueness or size of Earth's atmosphere. The Moon will be confined to a barren portion of the constellation Aries, far from any bright stars or naked-eye planets.

Table 8.17 **Contact Times: 2004 October 28**

Type	P1	U1	U2	Max	U3	U4	P2	UMag
Total	00:06	01:15	02:24	03:04	03:45	04:54	06:03	1.31

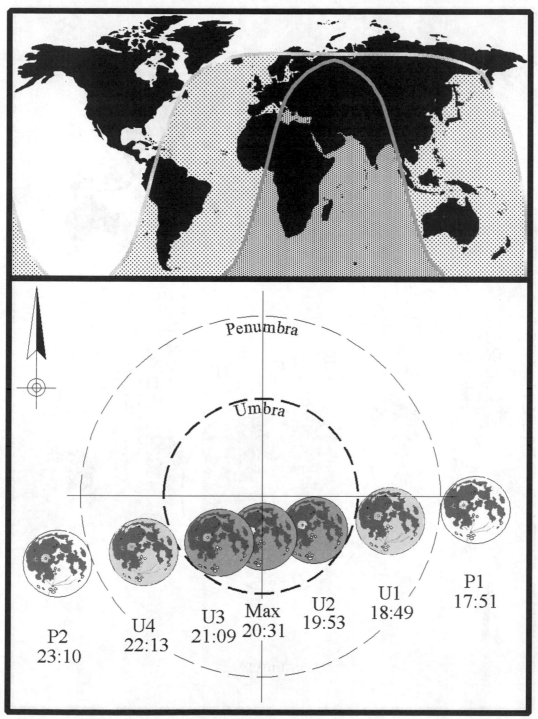

Figure 8.6 Circumstances of the total lunar eclipse of 2004 May 4.

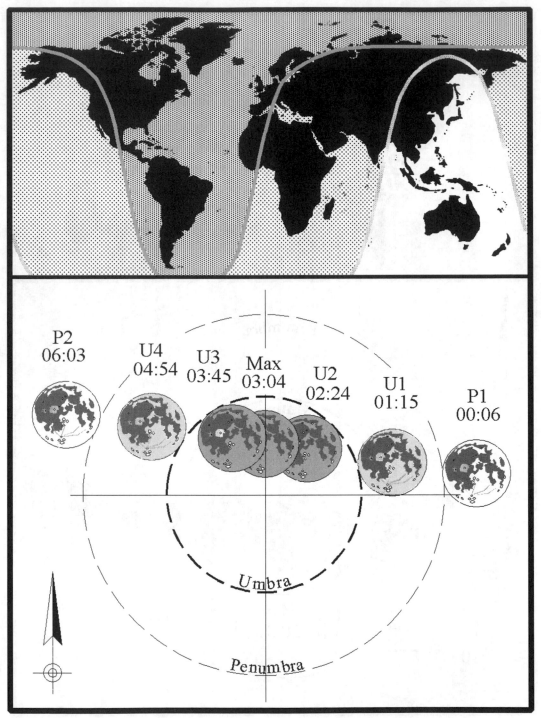

Figure 8.7 Circumstances of the total lunar eclipse of 2004 October 28.

PENUMBRAL LUNAR ECLIPSE OF 2005 APRIL 24

Only the western portions of North America, all of New Zealand, and the east coast of Australia will see this eclipse of the Moon. The best time to look will be between 09:30 and 10:20 U.T., when a slight charcoal shading may be visible along the Moon's northern edge. You'll find the Moon to the southeast of the bright star Spica in Virgo.

Table 8.18 **Contact Times: 2005 April 24**

Type	P1	U1	U2	Max	U3	U4	P2	PMag
Penumbral	07:50	—	—	09:55	—	—	12:00	0.89

PARTIAL LUNAR ECLIPSE OF 2005 OCTOBER 17

A minor partial umbral eclipse of the Moon, lasting just over an hour, will be visible this night across Alaska, Hawaii, western Canada, the northwest corner of the continental United States, New Zealand, and most of Australia. Even at maximum, the umbral magnitude will reach only 0.07, with the shadow covering only a minuscule portion of the southern lunar disk. Moonset will occur before the eclipse ends from the central United States and Canada.

Table 8.19 **Contact Times: 2005 October 17**

Type	P1	U1	U2	Max	U3	U4	P2	UMag
Partial	09:52	11:34	—	12:04	—	12:33	14:16	0.07

PENUMBRAL LUNAR ECLIPSE OF 2006 MARCH 14

The first eclipse of 2006 is a total event, albeit a total *penumbral* eclipse. Even though the Moon barely misses an entrance into the Earth's umbra, shading should be evident twenty to thirty minutes before and after maximum eclipse at 23:47 U.T. Observers in Europe and Africa are best situated to see the shadow's grayish hue along the Moon's southern limb, while anyone in the eastern third of North America and nearly all of South America should also spot evidence of the penumbral eclipse shortly after the Moon rises that evening.

Table 8.20 **Contact Times: 2006 March 14**

Type	P1	U1	U2	Max	U3	U4	P2	PMag
Penumbral	21:23	—	—	23:47	—	—	02:10	1.03

PARTIAL LUNAR ECLIPSE OF 2006 SEPTEMBER 7

A partial event will round out the year's lunar eclipses. The entire eclipse will be visible from western Australia, central Asia, and the eastern half of Africa. Europe will see the Moon rise in eclipse, while eastern Australia and New Zealand will watch the Moon set (on the morning of September 8) still in partial eclipse. At maximum, less than 20 percent of the Moon will be covered by Earth's umbra, with the northern lunar limb dipping deepest.

Table 8.21 **Contact Times: 2006 September 7**

Type	P1	U1	U2	Max	U3	U4	P2	UMag
Partial	16:43	18:06	—	18:52	—	19:38	21:01	0.19

TOTAL LUNAR ECLIPSE OF 2007 MARCH 3

Observers in Europe, Africa, and western Asia will be treated to a midnight total eclipse as the Moon takes over an hour to pass through Earth's umbra (figure 8.8). The eclipse will be an early-evening event for much of North and South America, the Moon rising already immersed in the shadow. Eastern Asia and even westernmost Australia will see a very early morning event.

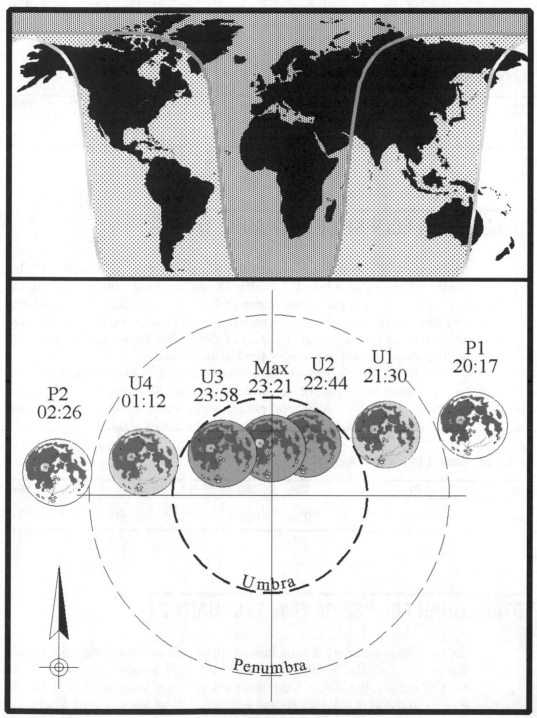

Figure 8.8 Circumstances of the total lunar eclipse of 2007 March 3.

The Moon will reside in the springtime constellation of Leo, framed by Regulus and Saturn to the northwest, and Spica, in Virgo, far to the southeast. At maximum, the eclipse will reach an umbral magnitude of 1.24.

Table 8.22 **Contact Times: 2007 March 3**

Type	P1	U1	U2	Max	U3	U4	P2	UMag
Total	20:17	21:30	22:44	23:21	23:58	01:12	02:26	1.24

TOTAL LUNAR ECLIPSE OF 2007 AUGUST 28

The year 2007 will bring a second total lunar eclipse on August 28. This event will be centered over the Pacific Ocean, favoring observers along the west coast of mainland United States and Canada as well as those in Alaska and Hawaii (figure 8.9). The entire eclipse will also be seen in easternmost Australia and New Zealand. The rest of the United States and Canada will see the Moon set still eclipsed by the Earth's umbra.

The Moon's track through the shadow will take it just south of the center. At maximum eclipse, the umbral magnitude will reach 1.48, potentially leading to a relatively dark total phase. At the time, the Moon will be situated in central Aquarius, far from any bright stars or planets.

Table 8.23 **Contact Times: 2007 August 28**

Type	P1	U1	U2	Max	U3	U4	P2	UMag
Total	07:53	08:51	09:52	10:38	11:23	12:24	13:23	1.48

TOTAL LUNAR ECLIPSE OF 2008 FEBRUARY 21

Six months later, another total lunar eclipse will be seen across the United States and Canada, South America, Africa, and Europe (figure 8.10). In North America, the Rocky Mountains will act as a dividing line. Those observers to the east will witness the entire event; to the west, the Moon will rise already in eclipse. Meanwhile, on the other side of the Atlantic, eclipse

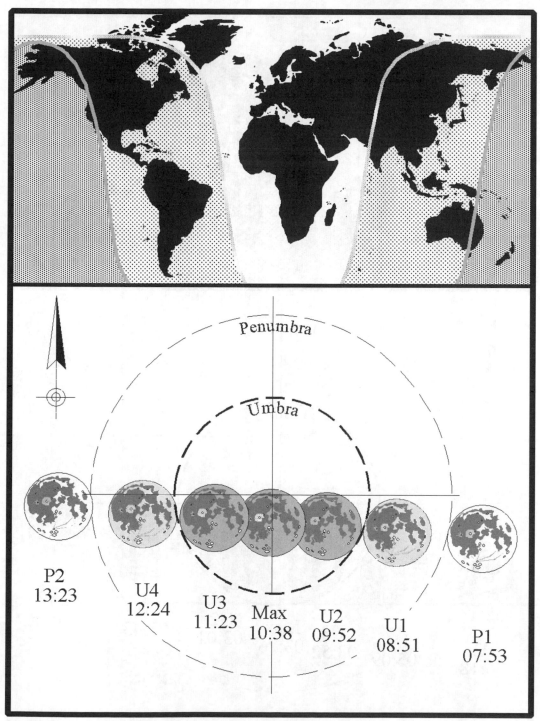

Figure 8.9 Circumstances of the total lunar eclipse of 2007 August 28.

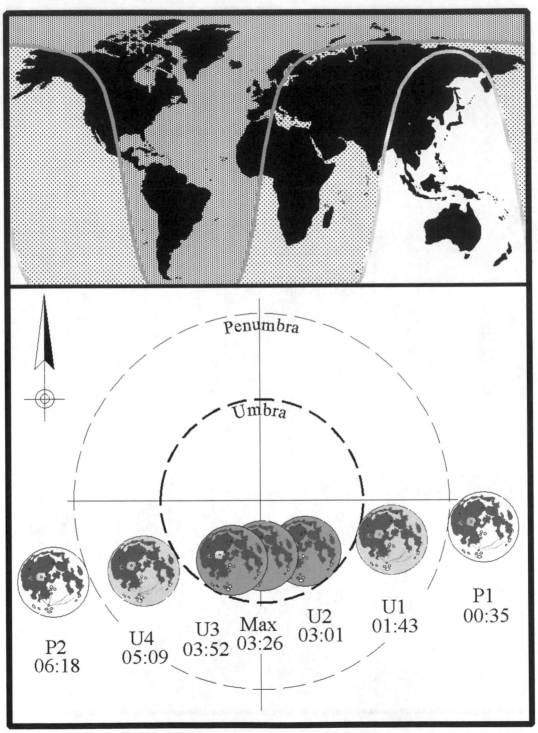

Figure 8.10 Circumstances of the total lunar eclipse of 2008 February 21.

watchers in western Europe will see the performance from start to finish, while anyone east of a north-south line connecting Italy and the Baltic Sea will watch the Moon set before the eclipse concludes.

At maximum eclipse, the Moon will just barely squeeze its entire disk into the southern portion of the umbra, leading to a potentially bright total phase highlighted by a brighter southern limb. The Moon will be stationed a few degrees to the southeast of the bright star Regulus in the constellation Leo. A second bright "star" will also be seen a little farther away, to the Moon's east-northeast. Inspection through a telescope with as little as 30-power will reveal that this is not a star at all, but rather the ringed planet Saturn.

Table 8.24 **Contact Times: 2008 February 21**

Type	P1	U1	U2	Max	U3	U4	P2	UMag
Total	00:35	01:43	03:01	03:26	03:52	05:09	06:18	1.11

PARTIAL LUNAR ECLIPSE OF 2008 AUGUST 16

More than three-quarters of the Moon will be covered by the Earth's umbra during this partial lunar eclipse. The entire event will be visible from eastern Europe, all of Asia, and most of Africa. Westernmost Europe and Africa, as well as South America, will see the Moon rise already in partial eclipse, while observers in Australia and New Zealand will watch the Moon set still shaded by Earth's shadow. Here in North America, the Moon will exit the shadow before or at moonrise, likely producing no observable effect. Maximum eclipse, when umbral immersion reaches 0.81 magnitude, occurs at 21:10 U.T.

Table 8.25 **Contact Times: 2008 August 21**

Type	P1	U1	U2	Max	U3	U4	P2	UMag
Partial	18:23	19:36	—	21:10	—	22:45	23:57	0.81

PENUMBRAL LUNAR ECLIPSE OF 2009 FEBRUARY 9

Although most penumbral eclipses produce little in the way of observable effects, this event will carry the Moon far enough into the light gray shadow to produce a dimming of the northern lunar limb. Eclipse devotees in Alaska, Hawaii, Australia, New Zealand, and eastern Asia should look around the time of maximum eclipse at 14:39 U.T. to see the greatest darkening.

Table 8.26　**Contact Times: 2009 February 9**

Type	P1	U1	U2	Max	U3	U4	P2	PMag
Penumbral	12:37	—	—	14:39	—	—	16:40	0.92

PENUMBRAL LUNAR ECLIPSE OF 2009 JULY 7

A pair of weak penumbral eclipses will occur during the summer of 2009. The Moon will be situated in the constellation Sagittarius and will remain above the horizon throughout the event for observers in North America west of the Great Lakes, including Hawaii and all but northernmost Alaska. The Moon will also be high in the sky over New Zealand and eastern Australia. There seems little hope, however, for seeing any dimming effect, since, at the point of maximum eclipse, the penumbral magnitude will reach only 0.16.

Table 8.27　**Contact Times: 2009 July 7**

Type	P1	U1	U2	Max	U3	U4	P2	PMag
Penumbral	08:38	—	—	09:38	—	—	10:39	0.16

PENUMBRAL LUNAR ECLIPSE OF 2009 AUGUST 6

The second summer lunar eclipse of 2009 will bring the Moon's northern limb into the Earth's penumbra, although again there seems little chance of seeing more than the slightest hint of this passage. For what there will be to see, sky watchers in Europe and Africa will be best placed to see it. The Moon will be in the constellation Capricornus during the eclipse, which, for them, will near the meridian at maximum.

Table 8.28 **Contact Times: 2009 August 6**

Type	P1	U1	U2	Max	U3	U4	P2	PMag
Penumbral	23:01	—	—	00:39	—	—	02:18	0.43

PARTIAL LUNAR ECLIPSE OF 2009 DECEMBER 31

While the rest of the world is preparing to celebrate the beginning of a new year, astronomers in Europe, Africa, Australia, and Asia will be celebrating the Moon's partial passage through the Earth's umbra. Actually, "celebrating" might be too strong a word for this partial eclipse. Even at maximum, umbral magnitude reaches only 0.08, with the Moon's southern limb barely tickling the Earth's central shadow. The Moon will be found that night in the constellation Gemini.

Table 8.29 **Contact Times: 2009 December 31**

Type	P1	U1	U2	Max	U3	U4	P2	PMag
Partial	17:16	18:53	—	19:23	—	19:54	21:31	0.08

PARTIAL LUNAR ECLIPSE OF 2010 JUNE 26

Only observers in Hawaii, western Alaska, Australia, New Zealand, and eastern portions of Malaysia and Asia stand much chance of seeing this partial lunar eclipse. Some evidence of the Moon's passage through the Earth's shadow may also be seen from the western edge of North America, although the Moon will set just as the eclipse is becoming interesting. Maximum occurs at 11:39 u.t., when the northern half of the lunar disk will be immersed in the umbra. You will find the Moon in the constellation Sagittarius, near the top of the familiar Teapot star pattern.

Table 8.30 **Contact Times: 2010 June 26**

Type	P1	U1	U2	Max	U3	U4	P2	UMag
Partial	08:56	10:17	—	11:39	—	13:01	14:22	0.54

TOTAL LUNAR ECLIPSE OF 2010 DECEMBER 21

The first total eclipse of the Moon in nearly three years will be visible to all of North America on this, the first day of winter (figure 8.11). Observers will find the Moon situated in Taurus, very close to the border with Gemini and near the most northerly point in the Moon's orbit. This also means that eclipse watchers in northern Europe and Asia will be able to see the event, even though it occurs during their "daytime." (At this time of year, from extreme northern latitudes, the Full Moon neither rises nor sets, but instead moves along the horizon.)

The path of the Moon will bring it just north of the umbra's center. From the track, it appears that this event should be moderately dark, yielding perhaps a brighter northern limb. The umbral magnitude reaches 1.26 at maximum.

Table 8.31 **Contact Times: 2010 December 21**

Type	P1	U1	U2	Max	U3	U4	P2	UMag
Total	05:28	06:33	07:41	08:17	08:54	10:02	11:06	1.26

TOTAL LUNAR ECLIPSE OF 2011 JUNE 15

Observers in Africa, southern Asia, and Australia may miss out on the total eclipse in December 2010, but they get their turn with this event (figure 8.12). This is liable to be one of the darkest eclipses listed in this book. At maximum, umbral magnitude reaches 1.71, second only to the 2000 July 16 total eclipse.

You will find the Moon within the confines of the constellation Ophiuchus the Serpent-Bearer, sometimes referred to, albeit unofficially, as the thirteenth sign of the zodiac. (The zodiac, or ecliptic—the apparent path of the Sun against the stars—actually spans a greater distance in Ophiuchus than in neighboring Scorpius.) What makes this a particularly interesting event is that, during totality, the Moon will occult the ninth-magnitude globular cluster NGC 6401. The best place to see the event will be in the northernmost region of the "entire-eclipse zone," in central Asia. There, as well as in eastern Europe and northeastern Africa, observers will see the Moon occult the cluster in a comparatively dark sky. Observers in western Europe will see the Moon rise already totally eclipsed and covering a por-

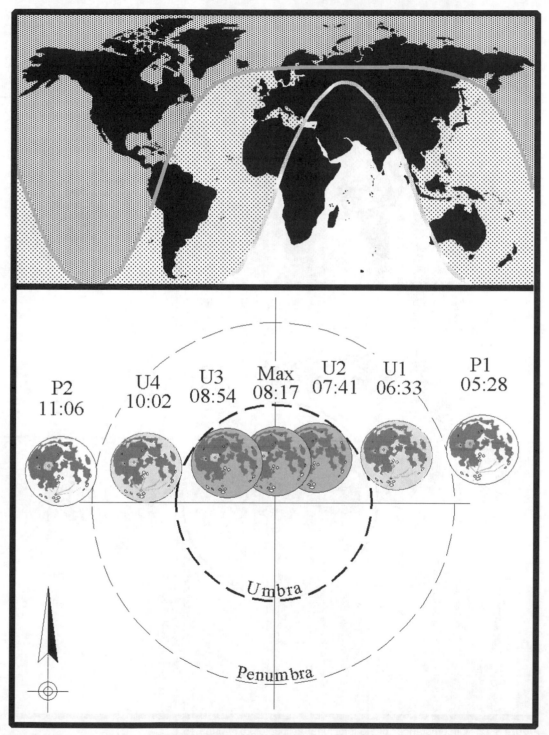

Figure 8.11 Circumstances of the total lunar eclipse of 2010 December 21.

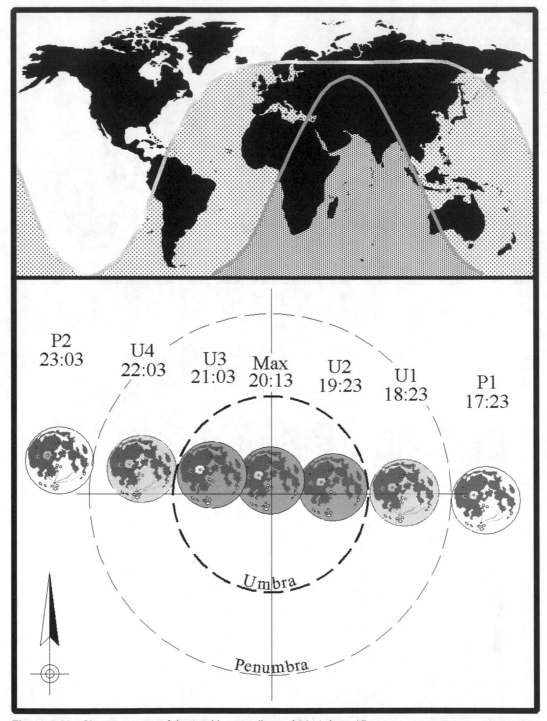

Figure 8.12   Circumstances of the total lunar eclipse of 2011 June 15.

tion of the cluster, while those in southern Africa and Australia will see the Moon slip past the cluster untouched. Note that you will need at least a 100-mm (4-inch) telescope to detect the cluster.

Table 8.32   **Contact Times: 2011 June 15**

Type	P1	U1	U2	Max	U3	U4	P2	UMag
Total	17:23	18:23	19:23	20:13	21:03	22:03	23:03	1.71

TOTAL LUNAR ECLIPSE OF 2011 DECEMBER 10

Alaska, northern Canada, Australia, New Zealand, and central and eastern Asia will have the best seats in the house for this next total lunar eclipse (figure 8.13). For everyone in Hawaii and the bulk of North America, the Moon will set still in eclipse. Those along the west coast of the United States and Canada will see the beginning of totality just as the Moon disappears below the western horizon; the rest of us will miss maximum eclipse. Indeed, East Coast residents will not even see the beginning of the umbral eclipse before moonset.

The eclipse will find the Moon set between the horns of Taurus. This will give photographers a great backdrop for some wonderfully creative wide-field pictures, with the ruddy lunar disk nestled among some of winter's brightest stars.

Table 8.33   **Contact Times: 2011 December 10**

Type	P1	U1	U2	Max	U3	U4	P2	UMag
Total	11:32	12:46	14:06	14:32	14:58	16:18	17:32	1.11

PARTIAL LUNAR ECLIPSE OF 2012 JUNE 4

A minor partial eclipse will be visible from observing sites within and surrounding the Pacific Ocean, including Hawaii, New Zealand, and central and eastern Australia. Observers in the western United States and Canada

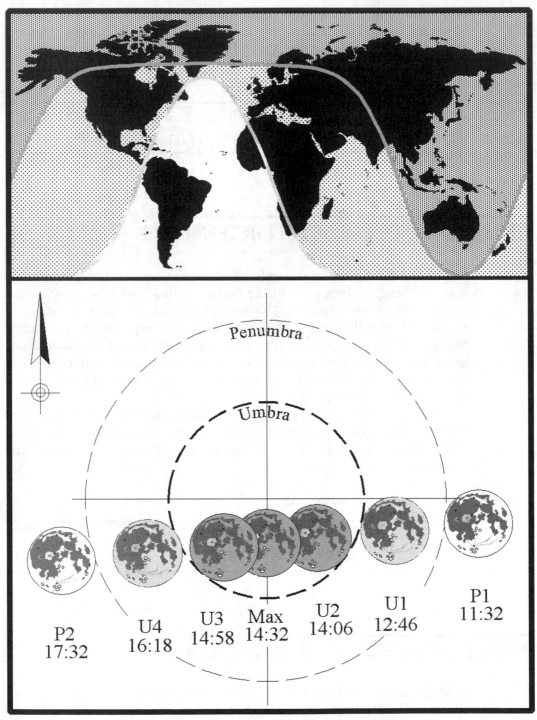

Figure 8.13 Circumstances of the total lunar eclipse of 2011 December 10.

will also be able to follow most of the event, although moonset will occur before the eclipse's conclusion. At maximum, the eclipse will attain an umbral magnitude of only 0.38, with the southern lunar limb deepest into the umbra. The Moon will be found in the constellation Ophiuchus, to the northeast of the bright red star Antares in neighboring Scorpius.

Table 8.34 **Contact Times: 2012 June 4**

Type	P1	U1	U2	Max	U3	U4	P2	UMag
Partial	08:47	10:00	—	11:04	—	12:08	13:20	0.38

PENUMBRAL LUNAR ECLIPSE OF 2012 NOVEMBER 28

This eclipse, with a maximum penumbral magnitude of 0.94, will be visible to observers in Alaska, Hawaii, New Zealand, Australia, and most of Asia. For perhaps 30 minutes on either side of the eclipse's maximum, a light gray shading will be noticeable along the Moon's northern limb. Although the eclipse itself will not be very impressive, the location of the Moon to the northwest of the Hyades in the constellation Taurus, and the bright planet Jupiter to the east-northeast, will create a pretty scene.

Table 8.35 **Contact Times: 2012 November 28**

Type	P1	U1	U2	Max	U3	U4	P2	UMag
Penumbral	12:13	—	—	14:33	—	—	16:54	0.94

PARTIAL LUNAR ECLIPSE OF 2013 APRIL 25

The Moon will barely touch the umbra during this very slight partial eclipse that will be visible across portions of central Africa and Asia. At greatest eclipse, the umbral magnitude will crest at a mere 0.02. The constellation Libra will be the site of the eclipse, with the Moon approximately 12.5° southeast of Virgo's brightest star, Spica. Another bright "star," not shown on most seasonal star charts, will be seen to the Moon's northeast. Inspection through a telescope will show this latter point to be the planet Saturn.

Table 8.36 **Contact Times: 2113 April 25**

Type	P1	U1	U2	Max	U3	U4	P2	UMag
Partial	18:03	19:52	—	20:08	—	20:25	22:14	0.02

PENUMBRAL LUNAR ECLIPSE OF 2013 MAY 25

A penumbral eclipse with a maximum magnitude of 0.04 is certainly not the stuff to excite observers into venturing outdoors! Yet, that's just what will occur over North and South America and western Europe and Africa during the Full Moon of 2013 May 25. Even at "maximum" eclipse (surely a misnomer in this case), there is no chance of seeing even the slightest hint of the eclipse as the Moon's southernmost limb just grazes the Earth's penumbra.

While the eclipse itself may be little more than a footnote in this book, tonight will feature an interesting pairing of the Full Moon and the star Graffias (Beta Scorpii) in the constellation Scorpius. In fact, the Moon will occult the star from the southeastern portion of the United States.

Table 8.37 **Contact Times: 2013 May 25**

Type	P1	U1	U2	Max	U3	U4	P2	PMag
Penumbral	03:45	—	—	04:11	—	—	04:37	0.04

PENUMBRAL LUNAR ECLIPSE OF 2013 OCTOBER 18

A deeper penumbral eclipse will await Moon watchers in Europe, Africa, and western Asia on 2013 October 18. Those of us along the east coast of the Americas might also notice a subtle darkening of the Moon's southern limb around moonrise, yet no matter where you are, the Moon's appearance will not change dramatically. Your best chance to catch a glimpse of the eclipse will be around maximum, 23:49 U.T. Throughout the event, the Moon will remain in a seemingly empty region of the autumn night sky occupied by Pisces.

Table 8.38 **Contact Times: 2013 October 18**

Type	P1	U1	U2	Max	U3	U4	P2	PMag
Penumbral	21:50	—	—	23:49	—	—	01:49	0.77

TOTAL LUNAR ECLIPSE OF 2014 APRIL 15

Amateur astronomers here in the United States had better not wait until the last minute to file their income taxes in 2014. If they do, they may end up working long into the night, missing the first total lunar eclipse widely visible from North America in four years (figure 8.14). On this night the Moon will slide silently through the Earth's umbra, passing just south of the umbra's center. The Moon will be completely eclipsed for 1 hour 19 minutes, with mid-eclipse occurring at 07:46 U.T. The Moon will set not long after totality ends from eastern North and South America, while central and western regions will see the entire event in dark skies.

The Moon will appear quite near the bright star Spica, in the constellation Virgo, during the eclipse. The farther north an observer is, the closer the Moon will be to the star, although not close enough for an occultation to take place. Even more enticing will be the bright red "star" to the Moon's northwest. But that's no star; that will be Mars. Mars will be close to opposition at the time of the eclipse, and ideally placed for observation of the Red Planet's surface features. On the day of the eclipse (and for several days on either side), Mars will measure 15 arc-seconds in diameter.

Table 8.39 **Contact Times: 2014 April 15**

Type	P1	U1	U2	Max	U3	U4	P2	UMag
Total	04:53	05:59	07:07	07:46	08:26	09:34	10:40	1.30

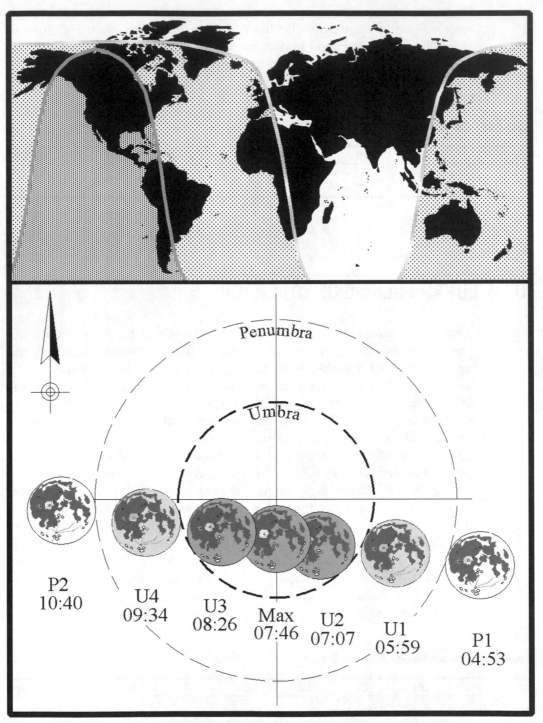

Figure 8.14 Circumstances of the total lunar eclipse of 2014 April 15.

TOTAL LUNAR ECLIPSE OF 2014 OCTOBER 8

This promises to be an exciting total lunar eclipse for observers in the central and western portions of North America, Hawaii, Alaska (especially Alaska; see below), and eastern Asia. Totality will last just under one hour from start to finish, the Moon passing to the north of the umbra's center (figure 8.15). As a result, we may expect a relatively bright eclipse, possibly featuring a coppery red tint, to the Moon's southern edge, contrasted by a brighter north polar region.

What makes this an especially interesting event, however, is the Moon's location in the sky within the constellation Pisces. While the constellation itself is devoid of bright stars, binoculars and telescopes will reveal a strange-looking sixth-magnitude greenish point of light directly in line with the lunar disk: the planet Uranus. It turns out that the planet and the Moon will appear closest to one another near the height of totality. Just how close they get depends on where an observer is located—the farther north, the better. From Los Angeles, for instance, the Moon will slide about 1° north of the planet's tiny disk, while from Vancouver the pair will be separated by half the distance. The most exciting view will come from central and northern Alaska and Canada (north of approximately 64° north latitude), where the Moon will actually occult Uranus during totality—a one-of-a-kind observation and photo opportunity, to be sure!

Table 8.40 **Contact Times: 2014 October 8**

Type	P1	U1	U2	Max	U3	U4	P2	UMag
Total	08:15	09:15	10:26	10:55	11:25	12:35	13:36	1.17

TOTAL LUNAR ECLIPSE OF 2015 APRIL 4

What might well be the brightest total lunar eclipse within this book's twenty-year span will take place on 2015 April 4 (that is, barring any "terrestrial" influences). That night the Moon will speed through the northernmost edge of Earth's umbra. With an umbral magnitude of 1.01, the entire lunar disk will just barely squeeze into the shadow during the 12 minutes of totality—the shortest duration since 1856 October 13. (For the record, an "instantaneous" total lunar eclipse occurred on 1899 December 17, al-

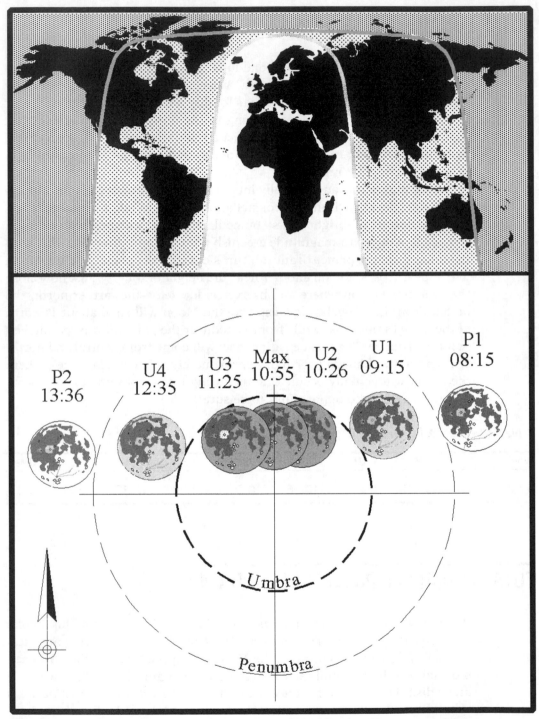

Figure 8.15 Circumstances of the total lunar eclipse of 2014 October 18.

though, since it had an umbral magnitude of 0.993, it should be classified as just barely a partial eclipse.)

Unfortunately, few people will see the event from start to finish, as the eclipse is centered over the Pacific Ocean (figure 8.16). Observers in western Alaska, Hawaii, New Zealand, and eastern Australia will have the best view, while those in the western continental United States and Canada will watch the Moon set still in eclipse. Moonset will occur in central and eastern North America, and all of South America, just as the partial umbral phases are getting under way. You will find the Moon in Virgo during this eclipse, northwest of the bright star Spica.

Table 8.41 **Contact Times: 2015 April 4**

Type	P1	U1	U2	Max	U3	U4	P2	UMag
Total	09:00	10:16	11:55	12:01	12:07	13:45	15:01	1.01

TOTAL LUNAR ECLIPSE OF 2015 SEPTEMBER 28

The final total lunar eclipse described in this book will occur with the Moon in the western portion of Pisces, directly below the prominent constellation of Pegasus the Flying Horse. Moon watchers in central and eastern North America (east of a line from roughly Houston, Texas, to Winnipeg, Manitoba) will see the complete eclipse, while those in the west will have to wait patiently for the Moon to rise after the eclipse has begun (figure 8.17). Across the Atlantic, astronomers in western Europe will also see the entire eclipse before moonset, although the Moon will disappear below the early-morning western horizon in eastern Europe and western Asia before its conclusion.

At maximum, the umbral magnitude tops out at 1.29, with the Moon passing south of the shadow's center. The overall appearance may be a moderately dark eclipse highlighted by a bright northern lunar limb.

Table 8.42 **Contact Times: 2015 September 28**

Type	P1	U1	U2	Max	U3	U4	P2	UMag
Total	00:11	01:08	02:11	02:48	03:25	04:28	05:25	1.29

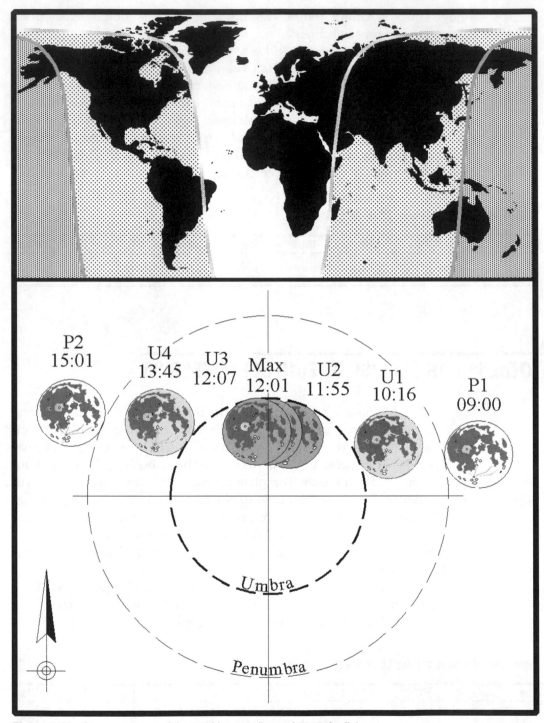

Figure 8.16 Circumstances of the total lunar eclipse of 2015 April 4.

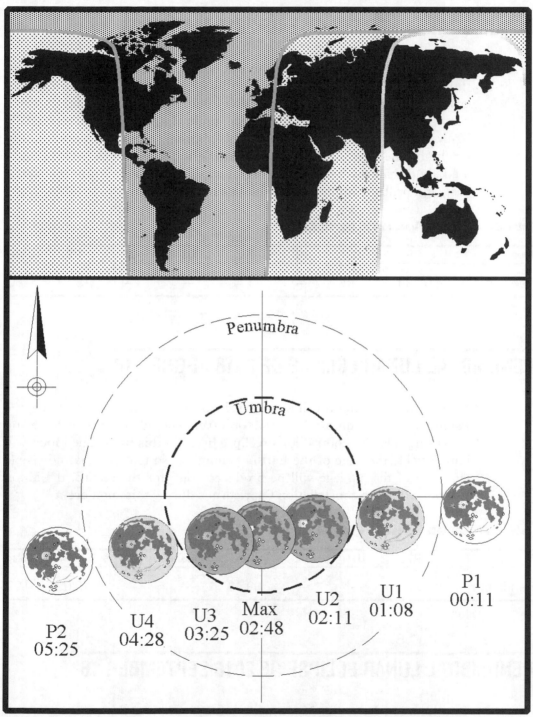

Figure 8.17 Circumstances of the total lunar eclipse of 2015 September 28.

PENUMBRAL LUNAR ECLIPSE OF 2016 MARCH 23

A relatively deep penumbral eclipse will be visible on 2016 March 23 to enthusiasts in Alaska, Hawaii, the far northwest coast of North America, eastern and central Australia, New Zealand, and easternmost Asia. At maximum, which occurs at 11:48 U.T., the penumbral magnitude will reach 0.80, likely producing a subtle but perceptible dimming along the Moon's southern limb. The Moon will be in the constellation Virgo for the eclipse, situated about halfway between the bright star Spica, about 20° to the southeast, and brilliant Jupiter, 20° to the Moon's northwest.

Table 8.43 **Contact Times: 2016 March 23**

Type	P1	U1	U2	Max	U3	U4	P2	UMag
Penumbral	09:38	—	—	11:48	—	—	13:58	0.80

PENUMBRAL LUNAR ECLIPSE OF 2016 AUGUST 18

Even though the Moon will be in the sky over western North America, the Pacific, and easternmost Australia, observers shouldn't get their hopes up for seeing this, the poorest lunar eclipse listed in this book. The Moon will barely tickle the edge of the Earth's penumbra, so tenuous that no visible evidence of the passage will be produced. But, for the record, the Moon will lie within the constellation Capricornus during this "non-event."

Table 8.44 **Contact Times: 2016 August 18**

Type	P1	U1	U2	Max	U3	U4	P2	UMag
Penumbral	09:25	—	—	09:43	—	—	10:02	0.02

PENUMBRAL LUNAR ECLIPSE OF 2016 SEPTEMBER 16

One lunation later, the Moon again bumps into Earth's penumbra, this time as seen from western Australia, Asia, and eastern Africa and Europe.

This crossing should be much more evident than the last, with the penumbral magnitude reaching 0.93 at 18:55 U.T., maximum eclipse. At that time a conspicuous dulling along the Moon's southern edge will be visible to the naked eye as well as through binoculars and telescopes. The Moon will be found in the northeastern corner of the constellation Aquarius, south of the more familiar Great Square of Pegasus.

Table 8.45 **Contact Times: 2016 September 16**

Type	P1	U1	U2	Max	U3	U4	P2	UMag
Penumbral	16:54	—	—	18:55	—	—	20:57	0.93

PENUMBRAL LUNAR ECLIPSE OF 2017 FEBRUARY 11

Sky watchers in New England and the Canadian Maritime provinces, along with their European and African counterparts, will enjoy a deep penumbral eclipse on 2017 February 11. Maximum eclipse will occur at 00:45 U.T., when the Moon's northern edge will just miss touching the Earth's darker umbra. You will find the Moon in western Leo, due west of Regulus, the brightest star in Leo's "sickle."

Table 8.46 **Contact Times: 2017 February 11**

Type	P1	U1	U2	Max	U3	U4	P2	UMag
Penumbral	22:33	—	—	00:45	—	—	02:56	1.01

PARTIAL LUNAR ECLIPSE OF 2017 AUGUST 7

I'm sorry the lunar-eclipse chapter can't end with the same impassioned excitement as chapter 7, but our last lunar event will be a comparatively unimpressive partial eclipse visible from Australia and central Asia. At maximum, umbral penetration will reach magnitude 0.25, with the Moon's southern quarter submerging in the umbra. Throughout, the Moon will be located in the dim constellation Capricornus.

Table 8.47 **Contact Times: 2017 August 7**

Type	P1	U1	U2	Max	U3	U4	P2	UMag
Partial	15:49	17:23	—	18:21	—	19:19	20:54	0.25

A wide variety of lunar eclipses await us as we forge into the next century. Some will attract little or no attention from the astronomical community, while others will be watched with great zeal and fervor. As with the solar eclipses outlined in the previous chapter, each lunar eclipse will feature a personality all its own.

Appendix A

Equipment Suppliers

Below are the addresses of all the manufacturers and suppliers listed in this book. Write to each for latest product information.

Telescopes

Astro-Physics, 11250 Forest Hills Road, Rockford, IL 61111; telephone (815) 282-1513

Celestron International, 2835 Columbia Street, Torrance, CA 90503; telephone (310) 328-9560

Edmund Scientific Company, N964 Edscorp Building, Barrington, NJ 08007; telephone (609) 573-6250

Meade Instruments Corporation, 16542 Millikan Avenue, Irvine, CA 92714; telephone (714) 756-2291

Orion Telescope Center, 2450 17th Avenue, P.O. Box 1158, Santa Cruz, CA 95061; telephone (800) 447-1001

Questar, P.O. Box 59, New Hope, PA 18938; telephone (215) 862-5277

Tele Vue Optics, 100 Route 59, Suffern, NY 10901; telephone (914) 357-9522

Solar filters

Orion Telescope Center, 2450 17th Avenue, P.O. Box 1158, Santa Cruz, CA 95061; telephone (800) 447-1001

DayStar Filter Corporation, P.O. Box 1290, Pomona, CA 91769; telephone (714) 591-4673

J.M.B., Inc., 20762 Richard, Trenton, MI 48183; telephone (313) 675-3490

Lumicon, 2111 Research Drive #5S, Livermore, CA 94550; telephone (510) 447-9570

Roger W. Tuthill, Inc., 11 Tanglewood Lane, Mountainside, NJ 07092; telephone (908) 232-1786

Thousand Oaks Optical, Box 5044, Thousand Oaks, CA 91359; telephone (800) 996-9111

Focusing Devices

Celestron International, 2835 Columbia Street, Torrance, CA 90503; telephone (310) 328-9560

P&S Sky Products, RR #1 20095 Con. 7, Mount Albert, Ontario, Canada L0G 1M0

Spectra-Astro-Systems, 6631 Wilbur Avenue, Suite 30, Reseda, CA 91335; telephone (800) 735-1352

Diffraction Gratings

Learning Technologies, Inc., 40 Cameron Avenue, Somerville, MA 02144; telephone (800) 537-8703

Edmund Scientific Company, N964 Edscorp Building, Barrington, NJ 08007; telephone (609) 573-6250

Satellite Signal Demodulators

Software Systems Consulting, 615 South El Camino Real, San Clemente, CA 92672; telephone (714) 498-5784

Photography Books

Kodak Color Print Viewing Filter Kit, Silver Pixel Press, 21 Jet View Drive, Rochester, NY 14624.

Topographic Maps

United States: USGS, Map Sales, Box 25286, Boulder, CO 80225 (in Alaska, write USGS, Map Sales, 101 12th Avenue, #12, Fairbanks, AK 99701)

Canada: Canada Map Office, 615 Booth Street, Ottawa, Ontario K1A 0E9

Australia: Division of National Mapping (NATMAP)

Similar maps are available from different agencies for other countries.

Computer Software

Not too long ago, predicting the exact times and location of an eclipse was a time-consuming task of mathematical "number crunching." But now, thanks to the personal computer and a whole universe of software, anyone can learn about a past or future eclipse in a matter of seconds. Here are a few that I used as references while writing this book, with the companies and individuals from whom they may be ordered.

Eclipse. Don Nicholson, 2124 Linda Flora Dr., Los Angeles, CA 90077.

Eclipse Complete. Charles Kluepfel, Zephyr Services, 1900 Murray Avenue, Pittsburgh, PA 15217; telephone (800) 533-6666

Lunar Eclipse. Christian Nuesch, Haldenstrasse 12, CH-8320 Fehraltorf, Switzerland; E-mail:christian.nuesch@squirrel.ch

Solar Eclipse. Matthew Merrill, 8927 Virginia Avenue, Apartment B, South Gate, CA 90280

The Sky. Software Bisque, 912 Twelfth Street, Suite A, Golden, CO 80401; telephone (800) 843-7599

Internet Sites

Listings of Astronomy Clubs

AstroNet:
http://www.rahul.resource/regular/clubs-etc/clubsetc.html

Astronomical Society of the Pacific:
http://maxwell.sfsu.edu/asp/amateur.html

Weather Information

Goddard Space Flight Center Weather Information:
http://climate.gsfc.nasa.gov

INTELLiCast Weather:
http://www.intellicast.com/weather

University of Michigan Weather Underground:
http://www.wunderground.com

Weather Channel:
http://www.weather.com

Travel Advisories

U.S. Department of State:
http://www.stolaf.edu/network/
travel-advisories.html

Solar and Lunar Eclipse Information

Goddard Space Flight Center Eclipse Prediction Information:
http://umbra.gsfc.nasa.gov/eclipse/
predictions/eclipse-paths.html

Solar Data Analysis Center, Solar Eclipse Information:
http://umbra.nascom.nasa.gov/eclipse

Solar and Lunar Eclipse Home Page (Fred Espenak, Goddard Space Flight Center):
http://planets.gsfc.nasa.gov/eclipse/
eclipse.html

Sky & Telescope's Eclipse Home Page:
http://www.skypub.com/eclipses/
eclipses.shtml

Five-Year Canon of Eclipses (Frank Roussel, Université de Rennes, France):
http://www.univ-rennes1.fr/ASTRO/
astro22e.html

Total Solar Eclipses: 1994–2010 (Catania Astrophysical Observatory, Italy):
http://www.ct.astro.it/eclissi.html

Eclipse Communications (Diane Kightlinger):
http://www.msen.com/~eclipse/

Eclipse Chaser Home Page (Jeffrey R. Charles, Versacorp):
http://www.eclipsechaser.com/

Lunar Eclipse Home Page (Anthony Mallama):
http://members.aol.com/LunarEcl/
lunar.htm

Lunar Eclipse Home Page (Byron Soulsby, Australia):
http://www.spirit.com.au/~minnah/
LEO.html

Lunar Occultations (International Occultation Timing Association):
http://www.sky.net/~robinson/iotandx.htm

Solar & Lunar Eclipse Simulator Software, PC Eclipse Program (Christian Nuesch, Switzerland):
http://www.astronomy.ch/

UMBRAPHILE, Macintosh Eclipse Program (Glenn Schneider, sofTouch Applications):
http://rtd.com/~gschneid/
UMBRAPHILE.html

Star Ware Home Page:
http://ourworld.compuserve.com/
homespages/pharrington

Appendix B

Bibliography and Further Reading

For readers who wish to continue their reading about solar and lunar eclipses, here is a list of reference books and periodicals that you might find interesting. Those with the notation "OP" are out of print, but may be available at larger public and university libraries as well as from antiquarian booksellers.

Astronomical Almanac. Washington, D.C.: U.S. Government Printing Office.

Astronomical Phenomena. Washington, D.C.: U.S. Government Printing Office.

Astronomy (periodical). Kalmbach Publishing, 21027 Crossroads Circle, P.O. Box 1612, Waukesha, WI 53187.

Astrophotography Basics (Kodak Publication Number P-150), Rochester, NY: Eastman Kodak.

Brewer, B. *Eclipse*. Seattle: Earth View, 1991.

Corliss, W. *Mysterious Universe: A Handbook of Astronomical Anomalies*. Glen Arm, MD: Sourcebook, 1979.

Espenak, F. *Fifty Year Canon of Lunar Eclipses* (NASA Reference Publication 1216); Cambridge, MA: Sky Publishing, 1989.

———. *Fifty Year Canon of Solar Eclipses* (NASA Reference Publication 1178 Revised); Cambridge, MA: Sky Publishing, 1994.

Graham, F., and J. Westfall. *Lunar Eclipse Handbook*. East Pittsburgh, PA: American Lunar Society, 1990.

Harris, J., and R. Talcott. *Chasing the Shadow*. Waukesha, WI: Kalmbach, 1994.

Journal of the Association of Lunar and Planetary Observers (periodical). P.O. Box 143, Heber Springs, AR 72543.

Littmann, M., and K. Willcox. *Totality: Eclipses of the Sun*. Honolulu, HI: University of Hawaii Press, 1991.

Meeus, J., C. Grosjean, and W. Vanderleen. *Canon of Solar Eclipses.* Oxford, England: Pergamon Press, 1966 (OP).

Mitchell, S. *Eclipses of the Sun.* New York, NY: Columbia University Press, 1924 (OP).

Ottewell, G. *The Under-Standing of Eclipses.* Greenville, SC: Furman University, 1991.

Pasachoff, J., and M. Covington. *The Cambridge Eclipse Photography Guide.* Cambridge, England: Cambridge University Press, 1993.

Phillips, K. J. H. *Guide to the Sun.* Cambridge, England: Cambridge University Press, 1992.

Reynolds, M., and R. Sweetsir. *Observe Eclipses.* Washington, D.C.: Astronomical League, 1995.

Selenology (periodical). American Lunar Society, P.O. Box 209, East Pittsburgh, PA 15112.

Sky & Telescope (periodical). Sky Publishing, P.O. Box 9111, Belmont, MA 02178.

Young, W., T. Benson, and G. Eaton. *Copying and Duplicating in Black-and-White and Color.* Rochester, NY: Eastman Kodak, 1984 (Silver Pixel Press, catalog no. 1527969).

Zimmermann, L. *Bad Astronomy: A Brief History of Bizarre Theories.* Piermont, NY: Linda Zimmermann, 1995.

Zirker, J. B. *Total Eclipses of the Sun.* New York: Van Nostrand Reinhold, 1995.

Appendix C

Societies and Associations

American Lunar Society, P.O. Box 209, East Pittsburgh, PA 15112

Association of Lunar and Planetary Observers, P.O. Box 143, Heber Springs, AR 72543

Astronomical League, 5027 West Stanford, Dallas, TX 75209

International Occultation Timing Association, 2760 S.W. Jewell Avenue, Topeka, KS 66611–1614; telephone (301) 464–4945

International Lunar Occultation Center, Geodesy and Geophysics Division, Hydrographic Department, Tsukiji-5, Chuo-ku, Tokyo 104, JAPAN

Eclipse Section, Flemish Astronomical Association, c/o Patrick Poitevin, Pulsebaan 16, 2275 Lille-Wechelderzande, BELGIUM

Appendix D

Solar Eclipse Tour Companies

This list of tour companies that sponsor solar eclipse expeditions is by no means complete, nor does it imply an endorsement. Consult the current issue of either *Astronomy* or *Sky & Telescope* for additional tour groups. In addition, contact local museums, planetaria, and astronomy clubs, as many sponsor their own eclipse tours.

Astronomical League eclipse tours, Ken Willcox, Route 2, P.O. Box 940, Bartlesville, OK 74006; telephone (800) 484-9073

Cruises Inc., 145 15th Street NE, Suite 403, Atlanta, GA 30309

Explorers Tours, 223 Coppermill Road, Wraysbury TW19 5NW, UNITED KINGDOM; telephone +44-1753 681999

Gayety Travel Services, 2115 Ralph Avenue, Brooklyn, NY 11234; telephone (800) 446-9328

International Expeditions, 1 Environmental Park, Helena, AL 35080; telephone (800) 633-4734

Outer Edge Expeditions, 4550 Pontiac Trail, Walled Lake, MI 48390; telephone (800) 322-5235

Scientific Expeditions, 227 West Miami Avenue, Suite #3, Venice, FL 34285; telephone (800) 344-6867

Travel Wizard, 7788 University Avenue, Le Mesa, CA 91941; telephone (800) 708-7677

Tropical Adventures, 2516 Sea Palm Drive, El Paso, TX 79936; telephone (800) 595-1003

Twilight Tours, 3316 West Chandler Boulevard, Burbank, CA; telephone (818) 841-8245

Appendix E

Converting Universal Time to Local Time

In order to standardize when an event is due to occur, eclipse contact times are listed by Universal Time (abbreviated u.t.) throughout this book. Universal Time (also known as Greenwich Mean Time or Coordinated Universal Time) is based on the local time at the Earth's prime meridian (0° longitude), which passes through Greenwich, England.

To find out when a celestial event is to take place from a particular location, its u.t. must be converted to the civil clock time. In general, there is one hour's difference for every 15° of longitude traveled away from the prime meridian. Figure AE.1 shows the Earth's many time zones, with conversion factors for each zone listed in the table at lower right. To convert u.t. to local time west of the prime meridian, subtract the value listed for a given zone from the Universal Time, remembering to change both the time and date if necessary. East of the prime meridian, *add* the appropriate value. If you are converting to daylight savings time, be sure to compensate to get the local time. (In other words, if you live west of the prime meridian, and must subtract 5 hours from u.t. to get your local standard time, subtract 4 hours to get the local Daylight Savings Time. East of the prime meridian, you would add one hour less to u.t.)

As an example, the total lunar eclipse of 2000 January 21 is due to begin at 3:22 u.t. An observer in the Eastern Standard Time (EST) zone of the United States will see the eclipse start at 10:22 P.M. on January 20, while the event will begin at 7:22 P.M. on January 20 for observers in Pacific Standard Time (PST).

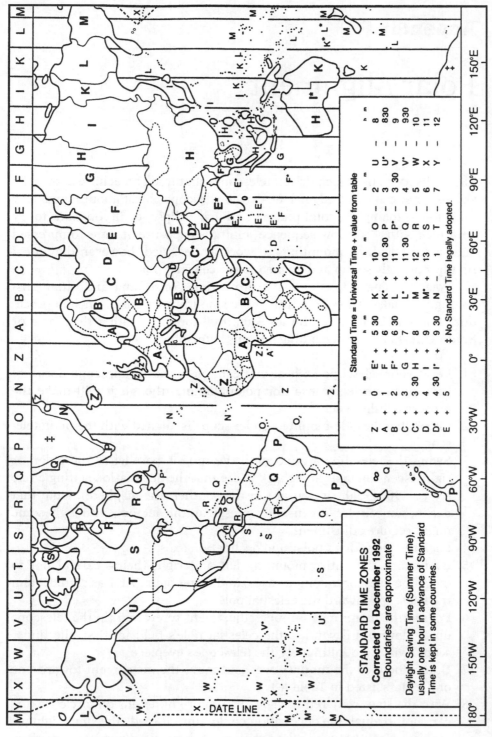

Figure AE.1 World Map of Standard Time Zones. Reproduced from *Astronomical Phenomena for the Year 1995* (Washington: U.S. Government Printing Office, and London: Her Majesty's Stationery Office, 1992).

Appendix F

Polar Alignment

Many observers use clock-driven telescopes that track an eclipse across the sky automatically. Indeed, a clock-driven telescope is just about mandatory for photographing the total phase of a lunar eclipse, or if you want to try your luck with a "shadow-sequence" multiple exposure or time-compressed videotape. For these mountings to work as designed, however, they must first be correctly set up and aligned to the celestial pole.

Polar-aligning a telescope can prove time-consuming and often frustrating. To make things go a little more smoothly, here is a ten-step procedure that can be used to aim an equatorial mount to either the North or South Celestial Pole (depending on your hemisphere).

1. Level your telescope as best you can. Note that a telescope does not have to be perfectly level for polar-aligning, though it will make the process a little easier.
2. Check that the side-mounted finder scope is aligned with the main telescope.
3. Swing the telescope around until the tube is parallel to the polar, or right-ascension, axis and check to see that the declination setting circle reads +90° (−90°, if you are in the southern hemisphere). Note that declination circles are usually preset at the factory, although some might require adjustment.
4. Lock the mounting's polar and declination axes.
5. By turning the entire mounting left and right (being careful not to move the telescope along its axes) and using figure AF.1 as a guide, aim the telescope toward the celestial pole.
6. Loosen the mounting's latitude adjustment screw and tip the telescope up or down until you see the celestial pole's field in, first, the finder scope's view and, ultimately, the telescope's eyepiece.
7. Unlock both of the mount's axes and move the instrument toward one of the stars listed in Table A.1.
8. With the stars centered in view, turn the right ascension circle (taking care not to touch the declination circle) until it reads the star's right ascension. Note that while the positions of these reference stars are given

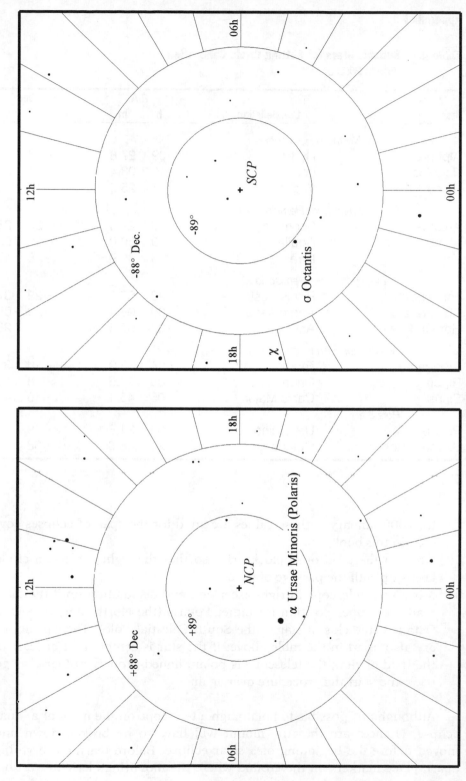

Figure AF.1 Use this chart when polar-aligning a telescope mount to help locate the exact positions of the North Celestial Pole (NCP, left) and South Celestial Pole (SCP, right).

Table A.1   **Suitable Stars for Setting-Circle Calibration**
(Epoch 2000.0)

Star	Constellation	R.A. h	R.A. m	Dec. °	Dec. '
Spring (Northern Hemisphere)					
Alphard	Hydra	09	27.6	−08	40
Regulus	Leo	10	08.4	+11	58
Spica	Virgo	13	25.2	−11	10
Summer (Northern Hemisphere)					
Antares	Scorpius	16	29.4	−26	26
Altair	Aquila	19	50.8	+08	52
Vega	Lyra	18	36.9	+38	47
Autumn (Northern Hemisphere)					
Fomalhaut	Piscis Austrinus	22	57.6	−29	37
Alpheratz	Andromeda	00	08.4	+29	05
Hamal	Aries	02	07.5	+23	28
Winter (Northern Hemisphere)					
Aldebaran	Taurus	04	35.9	+16	31
Rigel	Orion	05	14.5	−08	12
Sirius	Canis Major	06	45.1	−16	43
Pole Stars					
Polaris	Ursa Minor	02	31.8	+89	16
Sigma Octantis	Octans	21	08.6	−88	57

for 2000 January 1, these values are good for the span of eclipses covered in this book.

9. Turn on the telescope's clock drive so that the right ascension circle keeps up with the passage of time.

10. Move your telescope in right ascension and declination until the dials read the proper position for either Polaris (the North Star) or Sigma Octantis (the closest star to the South Celestial Pole). Their positions are also listed in the table above. If the star is centered, or at least in the field of view, the telescope is polar-aligned properly; otherwise, go back and start the procedure over again.

Although it is possible to polar-align a telescope on the night of a lunar eclipse, chances are that the mount will have to be broken down and moved before the beginning of a solar eclipse. Before tearing everything apart, place markers on the ground where the mounting's legs are resting.

This way, on the day of the eclipse, you can simply place the mounting back on the markers and—*voilá*—the telescope is (roughly) polar-aligned.

If the mounting doesn't track the sky properly even after it is properly polar-aligned, remember to check your hemisphere. Many of today's clock-driven telescopes feature drives that can be set for either northern- or southern-hemisphere operation. Make sure the telescope knows where it is!

What can be done if you cannot visit the observing site beforehand, as is often the case when you are part of a solar-eclipse expedition? How can an equatorial mount be aligned to the celestial pole? That's a tough one, but here's a method that worked pretty well for me during the July 1991 solar eclipse in Mexico. It should also work for you, provided you do it fairly close to the observing site (say, within 80 kilometers or so). First, set up and level your equatorial mounting, and proceed to polar-align it as you would normally. When the mounting is aligned, tighten everything down so that nothing will slip when it is disassembled. Before taking the mounting apart for transport, glue, tape, or otherwise affix a navigational compass on the mounting so that its arrow is pointed northward (or southward, if in the southern hemisphere). (Note that the compass reading will be only an approximation, since the Earth's magnetic poles are not coincident with the celestial poles.) When you finally get to the observing site on the day of the eclipse, you need only level the mounting and aim it so that the compass points in the same direction, and the mounting should be aimed at, or at least near, the celestial pole.

Appendix G

NASA Request Form for Solar Eclipse Bulletins

NASA Eclipse Bulletins contain detailed predictions, maps and meteorology for future central solar eclipses of interest. Published as part of NASA's Reference Publication (RP) series, the bulletins are prepared in cooperation with the Working Group on Eclipses of the International Astronomical Union, and are provided as a public service to both the professional and lay communities, including educators and the media. To allow a reasonable lead time for planning purposes, subsequent bulletins will be published 24 months or more before each event. Comments, suggestions, and corrections are solicited to improve the content and layout in subsequent editions of this publication series.

Single copies of the bulletins are available at no cost and may be ordered by sending the following form with a 9×12 inch (23×31 centimeter) SASE (self-addressed stamped envelope) with sufficient postage for each bulletin (11 ounces or 310 grams). Use stamps only, since cash or checks cannot be accepted. Requests within the United States may use the Postal Service's Priority Mail. Please print either the NASA RP number or the eclipse date (year and month) of the bulletin ordered in the lower left corner of the SASE, and return it with this completed form to either of the bulletin's authors, listed below. Requests from outside the United States and Canada may use international postal coupons to cover postage. Exceptions to the postage requirements will be made for international requests where political or economic restraints prevent the transfer of funds to other countries.

Request for: NASA Eclipse Bulletin for _____

Name of Organization: _____
(in English, if necessary)

Name of Contact Person: _____

Address: _____

_____

City/State/ZIP: _____

Country: _____

Type of organization: _____ University/College   _____ Observatory
(check all that apply)
                          _____ Library   _____ Planetarium

                          _____ Publication   _____ Media

                          _____ Professional   _____ Amateur

                          _____ Individual

Size of Organization: _____ (Number of Members)

Activities: _____

_____

_____

Return requests and comments to either of the following parties:

Fred Espenak   Jay Anderson
NASA/Goddard Space Flight Center   Manitoba Weather Services Centre
Planetary Systems Branch, Code 693   123 Main Street, Suite 150
Greenbelt, MD 20771 USA   Winnipeg, Manitoba, Canada R3C 4W2
Fax: (301) 286-0212   Internet: jander@cc.umanitoba.ca
Internet: u32fe@lepvax.gsfc.nasa.gov

Notes

1. The Dynamic Duo

1. Merletti, R., "The Sun's Obscuration at an Eclipse," "Astronomical Computing" column, *Sky & Telescope* (November 1986), 515–16.
2. van den Bergh, G., *Periodicity and Variation of Solar (and Lunar) Eclipses* (Haarlem, Netherlands: H. D. Tjeenk Willink, 1955).

2. An Eclipse Watcher's Shopping List

1. For readers who may be unfamiliar with telescope terminology, it is common practice to refer to a telescope by its aperture and focal ratio, rather than by overall focal length, as with camera lenses. So, a telescope referred to as a "20-cm (8-inch) f/10" has a mirror or front objective lens that measures 20 centimeters across and a focal length of 200 centimeters—or 2000 millimeters, about 80 inches—calculated by multiplying the aperture by the f/10 focal ratio.

3. Sun Worshiping

1. Author uncredited, *Total Eclipse Along the Eastern Seaboard*, *Sky & Telescope* (May 1970), 287.
2. di Cicco, D., "Trekking to Eclipse Magic," *Sky & Telescope* (July 1988), 100.
3. Ehmann, J. "Solar-Eclipse Sociology," *Sky & Telescope* (July 1988), 4.

4. A Bit of Luna-See

1. "October's Lunar Eclipse," *Sky & Telescope* (January 1988), 111.
2. Ashbrook, J., "The February Eclipse of the Moon II," *Sky & Telescope* (May 1971), 273.
3. Graham, F., and J. Westfall, *Lunar Eclipse Handbook* (Lunar Press, 1990). Authors Francis Graham and John Westfall recount how, through much trial and error, the Association of Lunar and Planetary Observers has devised the following formula that can be used to determine *an approxima-*

tion of the binocular correction factor:

$$F = 5 \log P + 0.3$$

F is the correction factor and P is the power, or magnification, of the binoculars. The constant (0.3) is used to account for some light lost in the optical system.

4. Reports include various articles published in *Sky & Telescope* magazine, as well as John Westfall, "Thirty Years of Lunar Eclipse Umbrae: 1956–1985" (*Journal of the Association of Lunar and Planetary Observers* 33, nos. 7–9, 112–17).

5. Eclipse Photography

1. A few words of caution about taking a shadow sequence: If you are using a German equatorial mount, make sure you will not have to "flip" the mount around, because the Moon will be crossing the meridian during the eclipse. This is one of the biggest drawbacks to this telescope-mounting design. As an object in view passes the meridian (the imaginary line passing from due north to due south, cutting the sky in half), a German equatorial mount is customarily swung around to make tracking easier. Doing this during a shadow sequence will prove disastrous, as the whole field of view will pivot and ruin the picture. Believe me, I know from experience!

7. Solar Eclipses: 1998–2017

1. *International Station Meteorological Climate Summary* (version 3.0), jointly produced by the Fleet Numerical Meteorology and Oceanography Detachment (United States Navy) and the National Climatic Data Center (National Oceanographic and Atmospheric Administration); available from the National Climatic Data Center, 151 Patton Avenue, Asheville, NC 28801.

Index